Lecture Notes in Computer Science 12530

More information about this subseries at http://www.springer.com/series/8379

Maciej Koutny · Fabrice Kordon ·
Lucia Pomello (Eds.)

Transactions on Petri Nets and Other Models of Concurrency XV

Springer

Editor-in-Chief
Maciej Koutny
Newcastle University
Newcastle upon Tyne, UK

Guest Editors
Fabrice Kordon ⓘ
Sorbonne Université
Paris, France

Lucia Pomello
Università degli Studi di Milano-Bicocca
Milan, Italy

ISSN 0302-9743 ISSN 1611-3349 (electronic)
Lecture Notes in Computer Science
ISSN 1867-7193 ISSN 1867-7746 (electronic)
Transactions on Petri Nets and Other Models of Concurrency
ISBN 978-3-662-63078-5 ISBN 978-3-662-63079-2 (eBook)
https://doi.org/10.1007/978-3-662-63079-2

This Springer imprint is published by the registered company Springer-Verlag GmbH, DE
part of Springer Nature.
The registered company address is: Heidelberger Platz 3, 14197 Berlin, Germany

Preface by Editor-in-Chief

The 15th issue of LNCS *Transactions on Petri Nets and Other Models of Concurrency* (ToPNoC) contains revised and extended versions of a selection of the best papers from the workshops held at the 40th International Conference on Application and Theory of Petri Nets and Concurrency (Petri Nets 2019, Aachen, Germany, 23–28 June 2019), and the 19th International Conference on Application of Concurrency to System Design (ACSD 2019, Aachen, Germany, 23–28 June 2019).

I would like to thank the two guest editors of this special issue: Fabrice Kordon and Lucia Pomello. Moreover, I would like to thank all authors, reviewers, and organizers of the Petri Nets 2019 and ACSD 2019 satellite workshops, without whom this issue of ToPNoC would not have been possible.

January 2021   Maciej Koutny

The original version of the book was revised: the affiliation of Lucia Pomello has been corrected. The correction to the book is available at
https://doi.org/10.1007/978-3-662-63079-2_9

LNCS Transactions on Petri Nets and Other Models of Concurrency: Aims and Scope

ToPNoC aims to publish papers from all areas of Petri nets and other models of concurrency ranging from theoretical work to tool support and industrial applications. The foundations of Petri nets were laid by the pioneering work of Carl Adam Petri and his colleagues in the early 1960s. Since then, a huge volume of material has been developed and published in journals and books as well as presented at workshops and conferences.

The annual International Conference on Application and Theory of Petri Nets and Concurrency started in 1980. For more information on the international Petri net community, see: http://www.informatik.uni-hamburg.de/TGI/PetriNets/.

All issues of ToPNoC are LNCS volumes. Hence they appear in all main libraries and are also accessible on SpringerLink (electronically). It is possible to subscribe to ToPNoC without subscribing to the rest of LNCS.

ToPNoC contains:

- Revised versions of a selection of the best papers from workshops and tutorials concerned with Petri nets and concurrency
- Special issues related to particular subareas (similar to those published in the *Advances in Petri Nets* series)
- Other papers invited for publication in ToPNoC
- Papers submitted directly to ToPNoC by their authors

Like all other journals, ToPNoC has an Editorial Board, which is responsible for the quality of the journal. The members of the board assist in the reviewing of papers submitted or invited for publication in ToPNoC. Moreover, they may make recommendations concerning collections of papers for special issues. The Editorial Board consists of prominent researchers within the Petri net community and in related fields.

Topics

The topics covered include: system design and verification using nets; analysis and synthesis; structure and behavior of nets; relationships between net theory and other approaches; causality/partial order theory of concurrency; net-based semantical, logical and algebraic calculi; symbolic net representation (graphical or textual); computer tools for nets; experience with using nets, case studies; educational issues related to nets; higher-level net models; timed and stochastic nets; and standardization of nets.

Also included are applications of nets to: biological systems; security systems; e-commerce and trading; embedded systems; environmental systems; flexible manu-facturing systems; hardware structures; health and medical systems; office automation;

operations research; performance evaluation; programming languages; protocols and networks; railway networks; real-time systems; supervisory control; telecommunications; cyber physical systems; and workflow.

For more information about ToPNoC see: http://www.springer.com/gp/computer-science/lncs/lncs-transactions/petri-nets-and-other-models-of-concurrency-topnoc-/731240

Submission of Manuscripts

Manuscripts should follow LNCS formatting guidelines, and should be submitted as PDF or zipped PostScript files to ToPNoC@ncl.ac.uk. All queries should be addressed to the same e-mail address.

Preface by Guest Editors

This volume of ToPNoC contains revised versions of a selection of the best workshop papers presented at satellite events of the 40th International Conference on Application and Theory of Petri Nets and Concurrency (Petri Nets 2019) and the 19th International Conference on Application of Concurrency to System Design (ACSD 2019). These events took place in Aachen, Germany in June 2019.

As guest editors, we are indebted to the program committees of the workshops and in particular to the chairs. Without their enthusiastic efforts, this volume would not have been possible.

The workshop papers considered for this special issue have been selected in close cooperation with the workshop chairs. Members of the program committees have participated in reviewing the new versions of the papers eventually submitted. We have received suggestions for papers for this special issue from:

- ATAED 2019: Workshop on Algorithms & Theories for the Analysis of Event Data (chairs: Wil M. P. van der Aalst, Robin Bergenthum, Josep Carmona),
- PNSE 2019: International Workshop on Petri Nets and Software Engineering (chairs: Ekkart Kindler, Daniel Moldt, Manuel Wimmer).

The authors of the suggested papers have been invited to improve and extend their results where possible, based on the comments received before and during the workshops. Each resulting revised submission was reviewed by at least two referees. We followed the principle of asking for fresh reviews of the revised papers, also from referees not involved initially in the reviewing of the original workshop contributions. All papers have gone through the standard two- or three-stage journal reviewing process, and eventually eight papers have been accepted after rigorous reviewing and revising.

Conformance checking, used to check if a log conforms to a model and vice versa, is a fundamental technique in process mining. The paper by Alessandro Berti and Wil M. P. van der Aalst, *A Novel Token-Based Replay Technique to Speed Up Conformance Checking and Process Enhancement*, presents an attempt to perform conformance checking exploiting a revisited token-based replay approach in contrast to alignment-based ones. The new proposed replay technique improves the speed and scalability of the algorithm, also for models with invisible transitions, and avoiding well-known problems such as token flooding.

Some applicative scenario often requires modeling a large set of variants of similar systems. In order to manage a unique model in these cases, the paper *Extensible Structural Analysis of Petri Net Product Lines*, by Elena Gómez-Martínez, Juan De Lara, and Esther Guerra, introduces the notion of Product Line Petri Net (PLPN), which represents a set of possible Petri nets. The authors show that certain structural properties of this set of Petri nets can be verified directly on the PLPN instead of checking them on each single model. This is done by reducing the problem to a

SAT-problem of some propositional formula. This approach is more efficient than checking explicitly each net of the product line.

It is known that the regional structure of a condition/event transition system, i.e. its set of regions with set inclusion and set complement, forms an orthomodular partial order (OMP). These regional structures are relevant to the relations between local states and local events of net systems and as such are of interest for the behavioural theory of Petri Nets. The paper *Stability of Regional Orthomodular Posets under Synchronisation and Refinement*, by Federica Adobbati, Carlo Ferigato, Stefano Gandelli, and Adrián Puerto Aubel, is a further step towards the characterization of orthomodular posets which are "stable", i.e. correspond to the regional structure of condition/event transition systems. The authors present a composition and a refinement operation for OMPs showing that both preserve stability.

Theoretical results in the area of synthesis of net models from behavioural specifications are presented in the paper *Efficient Synthesis of Weighted Marked Graphs with Circular Reachability Graph, and Beyond*, by Raymond Devillers, Evgeny Erofeev, and Thomas Hujsa. The authors specialize previous methods of synthesis of Conflict-Free (CF) nets and their Weighted Marked Graphs (WMG) subclass, two classes of weighted Petri nets allowing modelling of various real-world applications. They define conditions for checking the existence of a WMG whose reachability graph is isomorphic to a given LTS forming a single cycle, and propose two polynomial-time synthesis algorithms. The problems in extending these results to CF net synthesis are also discussed.

Another theoretical result dealing with Synthesis goes next. The paper *The Complexity of Synthesizing nop-Equipped Boolean Petri Nets from g-Bounded Inputs*, by Ronny Tredup, shows that computational complexity of π-synthesis remains hard for Boolean Petri nets g-bounded inputs, even when g is low (g being the maximum number of incoming and outgoing arcs). When g becomes very small, it sometimes makes the difference between hardness and tractability.

The next paper, *A Two-Player Asynchronous Game on Fully Observable Petri Nets*, by Federica Adobbati, Luca Bernardinello, and Lucia Pomello, deals with an asynchronous way to execute Petri nets. A Petri net is distributed if its elements can be assigned to a set of locations so that each element belongs to exactly one location, and each transition belongs to the same location as its input places. The paper defines an asynchronous game played on the unfolding of a distributed net with two locations, the "user" and the "environment". There are interesting applications of this work in situations where the "user" has to avoid dangerous situations that may occur in a system controlled by the "environment".

Model checking has to fight combinatorial explosion, especially in distributed (and thus asynchronous) systems. The paper *Solving finite-linear-path CTL-Formulas using the CEGAR Approach*, by Torsten Liebke and Karsten Wolf proposes a verification technique helping to solve CTL queries using the Petri net state equation with a CEGAR (counterexample guided abstraction refinement) approach. Such techniques are being implemented in the well-known LoLA 2 tool.

We end this special issue with *Verification of the MQTT IoT Protocol using Property-specific CTL Sweep-Line Algorithms*, by Alejandro Rodríguez, Lars Michael Kristensen, and Adrian Rutle. This paper is a case study dealing with MQTT, a publish-subscribe communication protocol increasingly being used for implementing internet-of-things (IoT) applications. This protocol is implemented using Coloured Petri Nets (CPNs) and behavioural properties are verified using CTL. To overcome combinatorial explosion during the process, the authors use the well known "sweep-line" method, which remains an open problem for CTL.

As guest editors, we would like to thank all the authors and referees who have contributed to this issue. The quality of this volume is the result of the high scientific value of their work. Moreover, we would like to acknowledge the excellent cooperation throughout the whole process that has made our work a pleasant task, despite the extremely challenging conditions our communities had to face in 2020 with the COVID pandemic. We are also grateful to the Springer/ToPNoC team for the final production of this issue.

January 2021

Fabrice Kordon
Lucia Pomello

Organization of This Issue

Guest Editors

Fabrice Kordon   Sorbonne Université, France
Lucia Pomello   University of Milano-Bicocca, Italy

Workshop Co-chairs

Wil M. P. van der Aalst   RWTH Aachen University, Germany
Robin Bergenthum   FernUniversität in Hagen, Germany
Josep Carmona   Universitat Politècnica de Catalunya, Spain
Ekkart Kindler   Technical University of Denmark, Denmark
Daniel Moldt   University of Hamburg, Germany
Manuel Wimmer   Johannes Kepler Universität Linz, Austria

Reviewers

Josep Carmona   Laure Petrucci
Evgeni Erofeev   Pascal Poizat
Stefan Haar   Adrián Puerto Aubel
Serge Haddad   Lorenzo Rossi
Ekkart Kindler   Stefan Schwoon
Jetty Kleijn   Yann Thierry-Mieg
Marta Koutny   Ronny Tredup
Lars Michael Kristensen   Karsten Wolf

Contents

A Novel Token-Based Replay Technique to Speed Up Conformance
Checking and Process Enhancement 1
    Alessandro Berti and Wil M. P. van der Aalst

Extensible Structural Analysis of Petri Net Product Lines. 27
    Elena Gómez-Martínez, Juan de Lara, and Esther Guerra

Stability of Regional Orthomodular Posets Under Synchronisation
and Refinement. 50
    Federica Adobbati, Carlo Ferigato, Stefano Gandelli,
    and Adrián Puerto Aubel

Efficient Synthesis of Weighted Marked Graphs with Circular Reachability
Graph, and Beyond. 75
    Raymond Devillers, Evgeny Erofeev, and Thomas Hujsa

The Complexity of Synthesizing *nop*-Equipped Boolean Petri Nets
from *g*-Bounded Inputs 101
    Ronny Tredup

A Two-Player Asynchronous Game on Fully Observable Petri Nets 126
    Federica Adobbati, Luca Bernardinello, and Lucia Pomello

Solving Finite-Linear-Path CTL-Formulas Using the CEGAR Approach 150
    Torsten Liebke and Karsten Wolf

Verification of the MQTT IoT Protocol Using Property-Specific CTL
Sweep-Line Algorithms 165
    Alejandro Rodríguez, Lars Michael Kristensen, and Adrian Rutle

Correction to: Transactions on Petri Nets and Other Models
of Concurrency XV. C1
    Maciej Koutny, Fabrice Kordon, and Lucia Pomello

Author Index 185

A Novel Token-Based Replay Technique to Speed Up Conformance Checking and Process Enhancement

Alessandro Berti[1,2]([⊠]) [iD] and Wil M. P. van der Aalst[1,2] [iD]

[1] Process and Data Science Group, RWTH Aachen University,
Lehrstuhl für Informatik 9, 52074 Aachen, Germany
a.berti@pads.rwth-aachen.de
[2] Fraunhofer Gesellschaft, Institute for Applied Information Technology (FIT),
Sankt Augustin, Germany

Abstract. Token-based replay used to be the standard way to conduct conformance checking. With the uptake of more advanced techniques (e.g., alignment based), token-based replay got abandoned. However, despite decomposition approaches and heuristics to speed-up computation, the more advanced conformance checking techniques have limited scalability, especially when traces get longer and process models more complex. This paper presents an improved token-based replay approach that is much faster and scalable. Moreover, the approach provides more accurate diagnostics that avoid known problems (e.g., "token flooding") and help to pinpoint compliance problems. The novel token-based replay technique has been implemented in the PM4Py process mining library. We will show that the replay technique outperforms state-of-the-art techniques in terms of speed and/or diagnostics.

Keywords: Log-model replay · Process diagnostics · Conformance checking

1 Introduction

The importance of conformance checking is growing as is illustrated by the new book on conformance checking [11] and the Gartner report which states "we see a significant trend toward more focus on conformance and enhancement process mining types" [14]. Conformance checking aims to compare an event log and a process model in order to discover deviations and obtain diagnostics information [25]. Deviations are related to process executions not following the process model (for example, the execution of some activities may be missing, or the activities are not happening in the correct order), and are usually associated to higher throughput times and lower quality levels. Hence, it is important to detect them, understand their causes and re-engineer the process in order to avoid such deviations. A prerequisite for both conformance checking and performance analysis is the ability to replay the event log on the model. This is

© Springer-Verlag GmbH Germany, part of Springer Nature 2021
M. Koutny et al. (Eds.): ToPNoC XV, LNCS 12530, pp. 1–26, 2021.
https://doi.org/10.1007/978-3-662-63079-2_1

needed to relate and compare the behavior observed in the log with the behavior observed in the model. Different replay techniques have been proposed, like *token-based replay* [27] and *alignments* [7,11]. In recent years, alignments have become the standard-de-facto technique since they are able to find an optimal match between the process model and a process execution contained in the event log. Unfortunately, their performance on complex process models and large event logs is poor. Token-based replay used to be the default technique for conformance checking, but has been almost abandoned in recent years, because the handling of invisible transitions and duplicate transitions require heuristics to select the proper path in the model. For example, models may get flooded with tokens in highly non-conforming executions, enabling unwanted parts of the process model and hampering the overall fitness evaluation. Moreover, more detailed diagnostics, that have been developed in recent years, have only been defined in the context of alignments.

In the paper [9], a revival of token-based replay is proposed. The approach improves the execution time of the token-based replay operation, increasing the performance gap between token-based replay and alignments (see Sect. 2.3). Moreover, the approach is able to manage the token-flood problem (see Sect. 3.3).

This contribution aims to extend [9] in some areas:

– *Root cause analysis* is introduced as a diagnostic (on the output of the token-based replay) provided by the approach.
– An analysis of a *backwards* state-space exploration approach (BTBR) is added. While this technique is not the main contribution of this paper, it provides a viable alternative to the state-of-the-art approach described in [27]. Moreover, some example applications are provided for both BTBR and ITBR.
– The evaluation section has been extended and includes now a detailed comparison of fitness values.

The remainder of the paper is organized as follows: in Sect. 2 an introduction to the main concepts used in this paper is provided. In Sect. 3, the problems are defined, and an improved token-based replay is proposed. In Sect. 4, some changes to the implementation are discussed and the evaluation of the approach is proposed. In Sect. 5, the tool support is presented, and we elaborate on the additional diagnostics (throughput time and root cause analysis). In Sect. 6, the related work is described. Section 7 concludes the paper.

2   Background

This section introduces standard concepts related to Petri nets and event logs. Moreover, the main token-based replay approach [27] is introduced.

2.1   Petri Nets

Petri nets provide a modeling language used from several process mining techniques, e.g., well-known process discovery[1] algorithms like the alpha miner and the inductive miner [16] (through conversion of the resulting process tree) can produce Petri nets. We start from the definition of elementary nets:

Definition 1 (Nets). *A net is a triple* (P, T, E) *where:*

- *P and T are disjoint sets of places and transitions respectively.*
- *$E \subseteq (P \times T) \cup (T \times P)$.*

Petri nets are such the set of arcs is a multiset over $(P \times T) \cup (T \times P)$.

Definition 2 (Multiset). *Let X be a set. $B \in \mathcal{B}(X) = X \to \mathbb{N}$ is a multiset over X where each element $x \in X$ appears $B(x)$ times. Between multisets $B_1 \in \mathcal{B}(X)$ and $B_2 \in \mathcal{B}(X)$ we define the following operations:*

- *(union) $B' = B_1 \cup B_2 \iff B'(x) = B_1(x) + B_2(x) \; \forall x \in X$. We can also say, in the same setting, that $B' = B_1 + B_2$*
- *(intersection) $B' = B_1 \cap B_2 \iff B'(x) = min(B_1(x), B_2(x)) \; \forall x \in X$.*
- *(multiset inclusion) $B_1 \leq B_2 \iff B_1(x) \leq B_2(x) \; \forall x \in X$. Conversely, $B_2 \geq B_1 \iff B_1 \leq B_2$.*
- *(difference) $B' = B_1 \setminus B_2 \iff B'(x) = max(B_1(x) - B_2(x), 0) \; \forall x \in X$. We can also say, in the same setting, that $B' = B_1 - B_2$.*

Moreover, we say that $x \in B \iff B(x) > 0$.

An accepting Petri net is a Petri net along with a final marking.

Definition 3 (Accepting Petri Nets). *A (labeled, marked) accepting Petri net is a net of the form $PN = (P, T, F, M_0, M_F, l)$ such that:*

- *P is the set of places.*
- *T is the set of transitions.*
- *$F \in \mathcal{B}((P \times T) \cup (T \times P))$ is a multiset of arcs.*
- *$M_0 \in \mathcal{B}(P)$ is the initial marking[2].*
- *$M_F \in \mathcal{B}(P)$ is the final marking.*
- *$l : T \to \Sigma \cup \{\tau\}$ is a labeling function that assigns to each transition $t \in T$ either a symbol from Σ (the set of labels) or the empty string τ.*

Definition 4 (Preset and Postset of a Place/Transition). *Let $x \in P \cup T$ be a place or a transition. Then $\bullet x, x \bullet \in \mathcal{B}(P \cup T)$ are defined such that:*

- *$\bullet x(y) = F((y, x)) \; \forall y \in P \cup T, (y, x) \in (P \times T) \cup (T \times P)$ is the preset of the element x.*

---

[1] With a process discovery technique, a process model is constructed capturing the behavior seen in an event log. See the book [4] for an introduction to the most popular process discovery algorithms.

[2] A marking $M \in \mathcal{B}(P)$ is a place multiset. We denote with $\mathcal{U}_M$ the universe of markings.

- $x \bullet (y) = F((x,y)) \; \forall y \in P \cup T, (x,y) \in (P \times T) \cup (T \times P)$ *is the postset of the element* x.

The initial marking corresponds to the initial state of a process execution. Process discovery algorithms may associate also a final marking to the Petri net, that is the state in which the process execution should end. A transition t is said to be *visible* if $l(t) \in \sum$; is said to be *invisible* if $l(t) = \tau$. If for all $t \in T$ such that $l(t) \neq \tau$, $|\{t' \in T | l(t') = l(t)\}| = 1$, then the Petri net contains *unique visible* transitions; otherwise, it contains *duplicate* transitions. In the following, some definitions in the context of nets and Petri nets are introduced.

Definition 5 (Execution Semantics). *The execution semantics of a Petri net is the following:*

- *A transition* $t \in T$ *is enabled (it may fire) in* M *if there are enough tokens in its input places for the consumptions to be possible, i.e. iff* $\bullet t \leq M$.
- *Firing a transition* $t \in T$ *in marking* M *produces the marking* $M' = (M \setminus \bullet t) \cup t \bullet$.

Definition 6 (Path). *A path of a net* $N = (P, T, E)$ *is a non-empty and finite sequence* $\eta_1, \ldots, \eta_n$ *of nodes of* $P \cup T$ *such that* $(\eta_1, \eta_2), \ldots, (\eta_{n-1}, \eta_n) \in E$. *A path* $\eta_1 \ldots \eta_n$ *leads from* η_1 *to* η_n.

Definition 7 (Strongly Connected Nets). *The net* $N = (P, T, E)$ *is strongly connected if a path exists between any node in* $P \cup T$, *i.e.,* $(x, y) \in E^* \; \forall \; x, y \in P \cup T$.

An important concept is the one of *structural components*, that is a collection of *subnets* with the property of holding at most one token per time during an execution of the net. Subnets, S-nets and S-components [13] are defined as follows.

Definition 8 (Subnets). $N' = (P', T', E')$ *is a subnet of* $N = (P, T, E)$ *if* $P' \subseteq P$, $T' \subseteq T$ *and* $E' = E \cap ((P' \times T') \cup (T' \times P'))$.

Definition 9 (S-nets). *A net* $N' = (P', T', E')$ *is an S-net if* $| \bullet t| = 1 = |t \bullet |$ *for every transition* $t \in T'$.

Definition 10 (S-Component). *A subnet* $N' = (P', T', E')$ *of* N *is an S-component of* N *if* $T' = \bullet P' \cup P' \bullet$ *and* N' *is a strongly connected S-net.*

2.2   Event Logs

In process mining, the definition of *event log* is fundamental, since it is the input of many techniques as process discovery and conformance checking.

Definition 1 (Event Log). *A log is a tuple* $L = (C_I, E, \Sigma, case\_ev, act, attr, \leq)$ *where:*

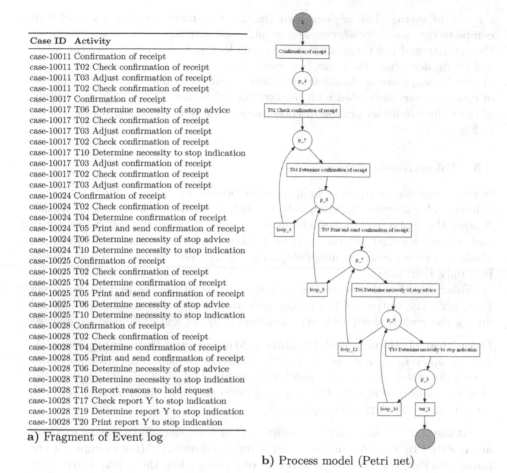

Case ID	Activity
case-10011	Confirmation of receipt
case-10011	T02 Check confirmation of receipt
case-10011	T03 Adjust confirmation of receipt
case-10011	T02 Check confirmation of receipt
case-10017	Confirmation of receipt
case-10017	T06 Determine necessity of stop advice
case-10017	T02 Check confirmation of receipt
case-10017	T03 Adjust confirmation of receipt
case-10017	T02 Check confirmation of receipt
case-10017	T10 Determine necessity to stop indication
case-10017	T03 Adjust confirmation of receipt
case-10017	T02 Check confirmation of receipt
case-10017	T03 Adjust confirmation of receipt
case-10024	Confirmation of receipt
case-10024	T02 Check confirmation of receipt
case-10024	T04 Determine confirmation of receipt
case-10024	T05 Print and send confirmation of receipt
case-10024	T06 Determine necessity of stop advice
case-10024	T10 Determine necessity to stop indication
case-10025	Confirmation of receipt
case-10025	T02 Check confirmation of receipt
case-10025	T04 Determine confirmation of receipt
case-10025	T05 Print and send confirmation of receipt
case-10025	T06 Determine necessity of stop advice
case-10025	T10 Determine necessity to stop indication
case-10028	Confirmation of receipt
case-10028	T02 Check confirmation of receipt
case-10028	T04 Determine confirmation of receipt
case-10028	T05 Print and send confirmation of receipt
case-10028	T06 Determine necessity of stop advice
case-10028	T10 Determine necessity to stop indication
case-10028	T16 Report reasons to hold request
case-10028	T17 Check report Y to stop indication
case-10028	T19 Determine report Y to stop indication
case-10028	T20 Print report Y to stop indication

a) Fragment of Event log

b) Process model (Petri net)

Fig. 1. Petri net extracted by the inductive miner on a filtered version of the "Receipt phase of an environmental permit application process" event log.

- C_I is a set of case identifiers.
- E is a set of events.
- Σ is the set of activities.
- $case\_ev \in C_I \rightarrow \mathcal{P}(E) \setminus \{\emptyset\}$ maps case identifiers onto set of events (belonging to the case).
- $act \in E \rightarrow \Sigma$ maps events onto activities.
- $attr \in E \rightarrow (\mathcal{U}_{attr} \nrightarrow \mathcal{U}_{val})$ (where $\mathcal{U}_{attr}$ is the universe of attribute names, and $\mathcal{U}_{val}$ is the universe of attribute values) maps events onto a partial function assigning values to some attributes.
- $\leq \ \subseteq E \times E$ defines a total order on events.

For a process supported by an information system, an event log is a set of cases, each one corresponding to a different execution of the process. A case contains the list of events that are executed (in the information system) in order to complete the case. To each case and event, some attributes can be assigned (e.g. the activity and the timestamp at the event level). A classification of the event is a string describing the event (e.g. the activity is a classification of the event). For each case, given a classification function, the corresponding trace is the list of classifications associated with the events of the case. An example application of the inductive miner process discovery algorithm to an event log is represented in Fig. 1.

2.3   Token-Based Replay

In process mining, a *replay* technique (as introduced in [4]) is a comparison of the behavior of a process execution with the behavior allowed by a process model. Among the replay techniques, the most important ones are *token-based replay* and *alignments* that act on Petri nets. Many different replay techniques are available in the process mining field, targeting different types of process models (not only Petri nets).

Token-based replay is applied to both a trace of the log and an accepting Petri net. The output of the replay operation is a list of transitions enabled during the replay, along with some numbers (c, p, m and r) defined as follows:

Definition 11 (Consumed, Produced, Missing, and Remaining Tokens).
Let L be an event log and σ be a trace of L. Then c is the number of consumed tokens during the replay of σ, p is the number of produced tokens during the replay of σ, m is the number of missing tokens during the replay of σ, and r is the number of remaining tokens during the replay of σ.

At the start of the replay, it is assumed that the tokens in the initial marking are inserted by the environment, increasing p accordingly (for example, if the initial marking consists of one token in one place, then the replay starts with $p = 1$). The replay operation considers, in order, the activities of the trace. In each step, the set of enabled transitions in the current marking is retrieved. If there is a transition corresponding to the current activity, then it is fired, a number of tokens equal to the sum of the input arcs is added to c, and a number of tokens equal to the sum of the output arcs is added to p. If there is not a transition corresponding to the current activity enabled in the current marking, then a transition in the model corresponding to the activity is searched (if there are duplicate corresponding transitions, then [27] provides an algorithm to choose between them). Since the transition could not fire in the current marking, the marking is modified by inserting the token(s) needed to enable it, and m is increased accordingly. At the end of the replay, if the final marking is reached, it is assumed that the environment consumes the tokens from the final marking, and c is increased accordingly. If the marking reached after the replay of the trace is different from the final marking, then missing tokens are inserted and remaining tokens r are set accordingly. The following relations hold during the replay: $c \le p + m$ and $m \le c$. The relation $p + m = c + r$ holds at the end of the replay. A fitness value can be defined for a trace and for the log.

Definition 12 (Trace Fitness). *Let L be an event log, σ be a trace of L, and c, p, m and r be the consumed, produced, missing and remaining tokens during the replay of σ. Then the fitness value for the trace is defined as:*

$$f_\sigma = \frac{1}{2}\left(1 - \frac{m}{c}\right) + \frac{1}{2}\left(1 - \frac{r}{p}\right)$$

Definition 13 (Log Fitness). *Let L be a log and $\sigma_0, \ldots, \sigma_n$ be the traces in L. Let $c(\sigma_i)$, $p(\sigma_i)$, $m(\sigma_i)$ and $r(\sigma_i)$ be the consumed, produced, missing and remaining tokens during the replay of trace σ_i. Then a fitness value for the log L is defined as:*

$$f_L = \frac{1}{2}\left(1 - \frac{\sum_{\sigma_i \in L} m(\sigma_i)}{\sum_{\sigma_i \in L} c(\sigma_i)}\right) + \frac{1}{2}\left(1 - \frac{\sum_{\sigma_i \in L} r(\sigma_i)}{\sum_{\sigma_i \in L} p(\sigma_i)}\right)$$

The log fitness is different from the average of fitness values at trace level. When, during the replay, a transition corresponding to the activity could not be enabled, and invisible transitions are present in the model, a technique is deployed to traverse the state space (see [27]) and possibly reach a marking in which the given transition is enabled. A heuristic (see [27]) that uses the shortest sequence of invisible transitions that enables a visible task is proposed. This heuristic tries to minimize the possibility that the execution of an invisible transition interferes with the future firing of another activity. A well-known problem for token-based replay is the *token flooding problem* [11]. Indeed, when the case differs much from the model, a lot of missing tokens are inserted during the replay. As a result of all the added tokens, many transitions become enabled. Therefore, also deviating events are likely to match an enabled transition. This leads to misleading diagnostics because unwanted parts of the model may be activated, and so the fitness value for highly problematic executions may be too high. To illustrate the token-flooding problem consider a process model without concurrency (only loops, sequences, and choices) represented as a Petri net. At any stage, there should be at most one token in the Petri net. However, each time there is a deviation, a token may be added, and that leads to a state which was never reachable from the initial state. The original token-based replay implementation [27] was only implemented in earlier versions of the ProM framework (ProM4 and ProM5) and proposes localized metrics on places of the Petri net that help to understand which parts of the model are more problematic. To improve performance in the original implementation, a preprocessing step is used to group cases having the same trace. In this way, multiple cases having the same trace only need to be analyzed once.

3   Approach

In Sect. 2.3, two problems were analyzed, that led to a relative abandonment of the token-based replay technique (older versions of ProM supported this, but ProM 6 does not):

1. The slowness in the traversal of invisible transitions, for which an expensive state-space exploration is required.
2. The token flooding problem.

To resolve the first problem, the methodology of exploration of the state-space needs to be changed. We describe in this section two approaches for token-based replay that address problem (1). In Sect. 3.1, an alternative approach (BTBR) to the one described in [27] is provided, in which a backwards state-space exploration is performed, instead of a forward state-space exploration. This leads to some advantages in managing common constructs in Petri net models, such as skips/loops, that are invisible transitions. However, a state-space exploration is still required and, with larger models, this is detrimental. After the backwards token-based replay, in Sect. 3.2 a novel technique (ITBR) is introduced. The improved token-based replay shows good performance results in the assessment.

Moreover, some approaches to solve problem (2) are proposed in Sect. 3.3, that exploit the properties of the process model in order to determine which tokens in the replay operation are useful and which are "superfluous".

3.1 Backwards Token-Based Replay

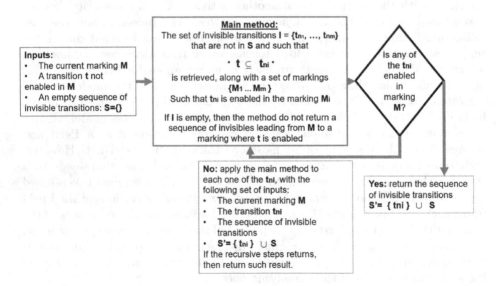

Fig. 2. A schema of the backwards activation algorithm for invisible transitions.

This section introduces an alternative token-based replay approach that is based on a *backwards* state-space exploration. The technique adopts the approach presented in [27] when a transition corresponding to the replayed activity is enabled

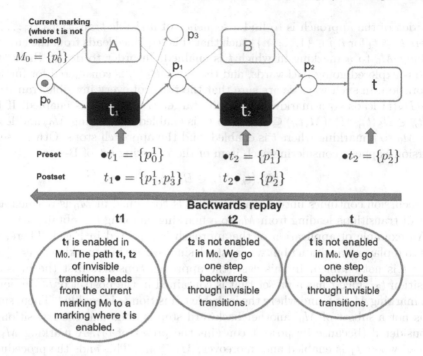

Fig. 3. An application the backwards activation approach for invisible transitions.

in the reached marking. When no corresponding transitions are enabled in the *current marking* M_0, an alternative approach (represented in Fig. 2) is followed to use invisible transitions and reach a marking where at least one corresponding transition is enabled. In the following, we suppose that $t \in T$ is a transition that corresponds to the replayed activity.

Definition 14 (Backwards Set of a Transition). *Let* $PN = (P, T, F, M_0, M_F, l)$ *be an accepting Petri net. We define the function:*

$$B_S : T \to \mathcal{P}(T)$$

$$B_S(t) = \{t' \in T \mid l(t') = \tau \ \wedge \ \bullet t \leq t' \bullet \}$$

Definition 15 (Backwards Marking). *Let* $PN = (P, T, F, M_0, M_F, l)$ *be an accepting Petri net. We define the function:*

$$M^{\leftarrow} : \mathcal{U}_M \times T \to \mathcal{U}_M$$

$$M^{\leftarrow}(M, t) = (M \setminus t \bullet) \cup \bullet t$$

As the backwards marking given M *and* t. *The backwards marking is such that* t *is enabled in* $M^{\leftarrow}(M, t)$.

The idea of the approach is to find a sequence of invisible transitions $t_1, \ldots, t_n$ (where $t_i \neq t_j$ for $i, j \in \{1, \ldots, n\}$) such that this sequence leads from the current marking M_0 to a marking in which t is enabled. In order to do so, the state-space is explored going backwards, and the B-set $B_S(t)$ is considered for further exploration. In such way, we are sure that the firing of every invisible transition $t_n \in B_S(t)$ leads to a marking where the target transition t is enabled. If for any $t_n \in B_S(t)$, $M^\leftarrow(M, t_n) \subseteq M_0$, then t_n is enabled in marking M_0, and leads from M_0 to a marking where t is enabled, and the approach stops. Otherwise, a recursion happens considering each item of the following set of B-sets

$$\{B_S(t_n) \mid t_n \in B_S(t)\}$$

The recursion continues until a marking that is contained in M_0 is reached and a list of transitions leading from M_0 to a marking enabling t is obtained.

An example of application of the approach is reported in Fig. 3. There, we need to replay a transition t but we are stuck since we are in a marking $M_0 = \{p_0^1\}$ where t is not enabled. In this case, the approach considers first the invisible transition t_2, since t has a preset that is contained in the postset of t_2. In doing so, a marking M_2 is found where the invisible transition t_2 is enabled. Then, since M_2 is not a subset of M_0, another backward step is done and the transition t_1 is considered (because its postset contains the preset of t_2). A marking M_1 is reached where t_1 is enabled and, moreover, $M_1 \subseteq M_0$. This ends the procedure, since from the current marking we are sure to be able to reach a marking where t is enabled by visiting transition t_1 and t_2: t_1 is enabled in M_0, t_2 is enabled by construction on the marking obtained firing t_1 on M_0, and t is enabled by construction on the marking obtained firing t_2.

The approach described in this section works nicely with models containing skip/loop transitions. Indeed, while the original token-based replay [27] needs to consider all the possibilities from the current marking, discarding some of them using heuristics, the backwards token-based replay approach considers the minimal marking in which a target transition is enabled and recursively explores the transitions which postset contains the minimal marking. However, the method is limited in the management of models with concurrency, given the B-set of a transition contains only the invisible transitions which completely enable that transition.

3.2   Improved Token-Based Replay

The method described in this part helps to enable a transition t through the traversal of invisible transitions. This helps to avoid the insertion of missing tokens when an activity needs to be replayed on the model, but no corresponding transition is enabled in the current marking M. Moreover, it helps to avoid time-consuming state-space explorations that are required by [27]. The approach works with accepting Petri nets in which the invisible transitions have non-empty preset and postset; this because any invisible transition with empty preset/postset would not belong to any shortest path between places. The description of the method starts from a preprocessing step on the Petri net, and continues with an algorithm to enable transitions using the results of this preprocessing step.

Preprocessing Step. Given an accepting Petri net $PN = (P, T, F, M_0, M_F, l)$, it is possible to define a directed graph $G = (V, A)$ such that the vertices V are the places P of the Petri net, and $A \subseteq P \times P$ is such that $(p_1, p_2) \in A$ if and only if at least one invisible transition connects p_1 to p_2. Then, to each arc $(p_1, p_2) \in A$, a transition $\tau(p_1, p_2)$ is associated to, picking one of the invisible transitions connecting p_1 to p_2.

Using an informed search algorithm for traversing the graph G, the shortest paths between nodes are found. These are sequences of places $\langle p_1, \ldots p_n \rangle$ such that $(p_i, p_{i+1}) \in A$ for any $1 \leq i < n$, and are transformed into sequences $\langle t_1, t_2, \ldots, t_{n-1} \rangle$ of transitions such that $t_i = \tau(p_i, p_{i+1})$ for any $1 \leq i < n$.

Given a marking M such that $M(p_1) > 0$ and $M(p_2) = 0$, a marking M' where $M'(p_2) > 0$ could[3] be reached by firing the sequence $\langle t_1, \ldots, t_n \rangle$ that is the shortest path in G between p_1 and p_2.

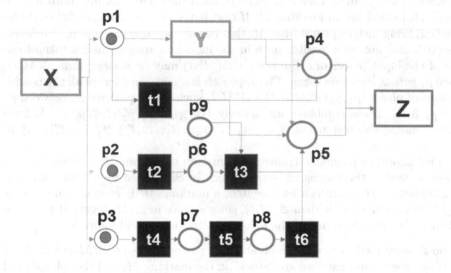

Fig. 4. A setting in which the application of the improved token-based replay is useful. The replay on this Petri net of the trace $\langle X, Z \rangle$ requires the firing of invisible transitions.

Enabling Transitions. This subsection explains how to apply the shortest paths to reach a marking where a transition is enabled. We start from defining the sets $\Delta(M, t)$ and $\Lambda(M, t)$.

Definition 16 (Delta Set and Lambda Set given a Marking and a Transition). *Let* $PN = (P, T, F, M_0, M_F, l)$ *be an accepting Petri net. Then we define:*

$$\Delta : \mathcal{U}_M \times T \to \mathcal{P}(P)$$

---

[3] During the activation of the sequence, some places could still have missing tokens.

$$\Delta(M, t) = \{p \in \bullet t \mid M(p) < F((p, t))\}$$

and

$$\Lambda : \mathcal{U}_M \times T \to \mathcal{P}(P)$$

$$\Lambda(M, t) = \{p \in P \mid F((p, t)) = 0 \land M(p) > 0\}$$

Given a marking M and a transition t, $\Delta(M, t)$ is the set of places that miss at least one token to enable transition t, while $\Lambda(M, t)$ is the set of places for which the marking has at least one token and t, in order to fire, does not require any of these places .

Given $\Delta(M, t)$ and $\Lambda(M, t)$, the idea is about using places in $\Lambda(M, t)$ (that are not useful to enable t) and, through the shortest paths, reach a marking M' where t is enabled.

Given a place $p_1 \in \Lambda(M, t)$ and a place $p_2 \in \Delta(M, t)$, if a path exists between p_1 and p_2 in G, then it is useful to see if the corresponding shortest path $\langle t_1, \ldots, t_n \rangle$ could fire in marking M. If that is the case, a marking M' could be reached, firing such sequence from M, that has at least one token in p_2. However, the path may not be a feasible path in the model, or may require a token from one of the input places of t. So, the set $\Delta(M', t)$ may be smaller than $\Delta(M, t)$, since p_2 gets at least one token. The approach is about considering all the combinations of places $(p_1, p_2) \in \Lambda(M, t) \times \Delta(M, t)$ such that a path exists between p_1 and p_2 in G. These combinations, namely $\{(p_1, p_2), (p'_1, p'_2), (p''_1, p''_2), \ldots\}$, have corresponding shortest paths $S = \{\langle t_1, \ldots, t_m \rangle, \langle t'_1, \ldots, t'_n \rangle, \langle t''_1, \ldots, t''_o \rangle, \ldots\}$ in G.

The algorithm to enable transition t through the traversal of invisible transitions considers the sequences of transitions in S, ordered by length, and tries to fire them. If the path can be executed, a marking M' is reached, and the set $\Delta(M', t)$ may be smaller than $\Delta(M, t)$, since a place in $\Delta(M, t)$ gets at least one token in M'. However, one of the following situations could happen:

1. no shortest path between combinations of places $(p_1, p_2) \in \Lambda(M, t) \times \Delta(M, t)$ could fire: in that case, we are "stuck" in the marking M, and the token-based replay is forced to insert the missing tokens;
2. a marking M' is reached, but $\Delta(M', t)$ is not empty, hence t is still not enabled in marking M'. In that case, the approach is iterated on the marking M';
3. a marking M' is reached, and $\Delta(M', t)$ is empty, so t is enabled in marking M'.

When situation (2) happens, the approach is iterated. A limit on the number of iterations may be set, and if it is exceeded, then the token-based replay inserts the missing tokens in marking M. The approach is straightforward when sound workflow nets without concurrency (only loops, sequences, and choices) are considered, since in the considered setting (M marking where transition t is not enabled) both sets $\Lambda(M, t)$ and $\Delta(M, t)$ have a single element, a single combination $(p_1, p_2) \in \Lambda(M, t) \times \Delta(M, t)$ exists and, if a path exists between p_1 and p_2 in G, and the shortest path could fire in marking M, a marking M' will be reached such that $\Delta(M', t) = \emptyset$ and the transition t is enabled.

Moreover, it performs particularly well on models that are output of popular process discovery algorithms (inductive miner [16], heuristics miner [30], ...) where potentially long chains of invisible (skip, loop) transitions need to be traversed in order to enable a transition. The approach described in this subsection can also manage duplicate transitions corresponding to the activity that needs to be replayed. In that case, we are looking to enable one of the transitions belonging to the set $T_C \subseteq T$ that contains all the transitions corresponding to the activities in the trace. The approach is then applied on the shortest paths between places. A similar approach can be applied to reach the final marking when, at the end of the replay of a trace, a marking M is reached that is not corresponding to the final marking. In that case, $\Delta = \{p \in P \mid M(p) < M_F(p)\}$ and $\Lambda = \{p \in P \mid M_F(p) = 0 \land M(p) > 0\}$. This does not cover the case where the reached marking contains the final marking but has too many tokens.

An example application of the approach is contained in Fig. 4. There, after executing X we have three tokens, one in p_1 one in p_2 and one in p_3. The next replayed activity is Z, that requires one token in p_4 and one token in p_5. However, since Y is not executed, both tokens are missing. From the marking $\{p_1, p_2, p_3\}$ to the set of missing tokens $\{p_4, p_5\}$, the set of shortest paths is $S = \{\langle p_1, p_4 \rangle, \langle p_2, p_6, p_5 \rangle, \langle p_3, p_7, p_8, p_5 \rangle\}$. These are ordered by the length of the path. Starting from the first path, transition t_1 is enabled, and p_4 is reached. Then, the second path is examined, however t_3 cannot fire hence p_5 cannot be reached. So, the last path is executed, and that leads to putting one token in p_5 and eventually enabling t.

3.3 Addressing the Token Flooding Problem

To address the token flooding problem, which is one of the most severe problems when using token-based replay, we propose several strategies. The final goal of these strategies is to avoid the activation of unwanted transitions that get enabled by the insertion of missing tokens, keeping the fitness value low for the problematic parts of the model. The common pattern behind these strategies is to determine *superfluous tokens*, that are tokens that cannot be used anymore. During the replay, f (initially set to 0) is an additional variable that stores the number of "frozen" tokens. When a token is detected as superfluous, it is "frozen": that means, it is removed from the marking and f is increased. Frozen tokens, like remaining tokens, are tokens that are produced in the replay but never consumed. Hence, at the end of the replay $p + m = c + r + f$. To each token in the marking, an *age* (number of iterations of the replay for which the token has been in the marking without being consumed) is assigned. The tokens with the highest age are the best candidates for removal. The techniques to detect superfluous tokens are deployed when a transition requires the insertion of missing tokens to fire, since the marking would then possibly contain more tokens. One of the following strategies can be used:

1. Using a decomposition of the Petri net in semi-positive invariants [18] or S-components [1,13] to restrict the set of allowed markings. Considering S-

components, each S-component should hold at most 1 token, so it is safe to remove the oldest tokens if they belong to a common S-component.

2. Using place bounds [20]: if a place is bounded to n tokens and during the replay operation the marking contains $m > n$ tokens for the place, the "oldest" tokens according to the age are removed.

4   Implementation and Evaluation of the Improved Token-Based Replay Technique

In this section, we present some changes to the implementation that have been performed in order to increase the performance of ITBR. Moreover, we present an assessment of ITBR on real-life logs.

4.1   Changes to the Implementation to Improve Performance

In our implementation of token-based replay, we adapt some ideas first used in the context of alignments [6]:

1. *Post-fix caching*: a post-fix is the final part of a case. During the replay of a case, the couple marking+post-fix is saved in a dictionary along with the list of transitions enabled from that point to reach the final marking of the model. For the next replayed cases, if one of them reaches exactly a marking + post-fix setting saved in the dictionary, the final part of the replay is retrieved from the dictionary.

2. *Activity caching*: the list of invisible transitions that are activated, from a given marking, to reach a marking where a particular transition is enabled, is saved into a dictionary. For the next replayed cases, if one of them reaches a marking + target transition setting saved in the dictionary, then the corresponding invisible transitions are fired accordingly to enable the target transition.

In the following:

- *CTBR* is the classical token-based replay (implemented in ProM 5).
- *ITBR* is the improved token-based replay described in this paper (implemented in PM4Py).
- *ABR* is the alignment-based replay (implemented in the "Replay a Log on Petri Net for Conformance Analysis" plug-in of ProM 6).
- *BTBR* is the token-based with backwards state-space exploration described in this paper (implemented in PM4Py).
- *AFA* is the approach described in [24] (implemented in PM4Py).
- *REABR* is the recomposition approach described in [15] (available in ProM 6).
- $ITBR^{-PC}$ is the improved token-based replay without postfix caching
- $ITBR^{-AC}$ is the improved token-based replay without activity caching
- $ITBR^{-PC/-AC}$ is the improved token-based replay without activity or postfix caching.

4.2   Evaluation: Execution Time

Table 1. Performance of the different replay approaches on real-life logs and models extracted by the inductive miner. The first columns contain some features of the log. The middle columns compare ITBR with ABR. In the rightmost columns, the performance of the BTBR, AFA and REABR approaches (BTBR was unable to analyze two of the datasets) are included.

Log	Cases	Variants	T.ITBR	T.ABR	T.BTBR	T.AFA	T.REABR
repairEx	1104	77	0.06	0.2	0.04	**0.03**	0.8
reviewing	100	96	**0.10**	0.4	0.29	0.11	2.2
bpic2017 (offer)	42995	16	0.30	1.5	0.06	**0.01**	0.18
receipt	1434	116	**0.09**	0.8	0.25	0.10	0.91
roadtraffic	150370	231	1.03	5.3		**0.09**	1.37
Billing	100000	1020	1.36	8.0	2.04	**1.18**	9.7
bpic2017 (application)	31509	15930	**56.1**	1520.3		116.7	1369.2
bpic2018	43809	28457	**145.8**	8427.2	400.99	543.01	2550.0
bpic2019	251734	11973	**27.0**	599.1	84.60	97.50	435.9

In this section, the improved token-based replay (ITBR) is assessed, looking at the speed and the output of the replay, against the alignment-based approach on Petri nets (ABR) and the other considered approaches. Tests contained in Table 1 are performed on real-life logs that can be retrieved from the 4TU log repository[4]. The tests have been done on an Intel I7-5500U powered computer with 16 GB DDR4 RAM.

Comparison Against ABR. For real-life logs and models extracted by the inductive miner, the ITBR is 5 times faster on average. Even for large logs, the replay time is less than a few seconds. For the latest BPI Challenge logs, given the model extracted by the inductive miner implementation in PM4Py, there is a noticeable speedup that is $> 20x$, but also the token-based replay is taking over 20 s.

ABR produces a different output than the one of token-based replay, so results are not directly comparable. Both are replay techniques, so the goal of both techniques is to provide information about fitness according to the process model (albeit the fitness measures are defined in a different way, and so are intrinsically different). This is valid in particular for the comparison of execution times: a trace may be judged fitting according to a process model in a significantly lower amount of time using token-based replay in comparison to alignments. If an execution is unfit according to the model, it can also be judged unfit in a significantly lower amount of time. For a comparison, read Section 8.4 of book [11] or consult [5,26].

---

[4] The logs are available at the URL https://data.4tu.nl/repository/collection:event_logs.

Comparison Against BTBR. The comparison between ITBR and BTBR shows that generally ITBR has significantly better performance on larger logs (BPI Challenge 2017 application, BPI Challenge 2018, BPI Challenge 2019). This shows that the preprocessing step helps to get better performance from token-based replay. The BTBR approach seems also limited in the type of process models it can handle: while it succeeds for 7 of the considered logs/models, it fails for two settings due to concurrency in the process model.

Comparison Against AFA. The alignments on finite automaton approach (AFA) shows better performance than ITBR for the vast majority of the logs, excluding the three bigger logs that have been considered (BPI Challenge 2017 application, BPI Challenge 2018, BPI Challenge 2019). Moreover, it shows significantly better performance than the other two evaluated alignments approaches (ABR and REABR) working on Petri nets. Possibly, the worse results of AFA against ITBR in the three BPI Challenge logs have been caused by the bigger size of the automaton, that can grow fast in complexity.

Comparison Against REABR. The alignments approach based on a maximal decomposition and, then, a recomposition of the results (REABR) shows a significant performance increase in comparison to classical alignments (ABR), showing the effectiveness of the approach. However, it records worse results than AFA that is performed on a different class of models (finite automatons) and ITBR.

Table 2. Comparison of the ITBR execution times on models extracted by the inductive miner on the given logs with or without postfix and activity caching. Here, the first column is the name of the log, the second is the execution time of ITBR without postfix and activity caching, the third is the execution time of ITBR without activity caching, the fourth is the execution time of ITBR without the postfix caching, the fifth is the execution time of ITBR with activity and postfix caching enabled.

Log	ITBR$^{-PC/-AC}$(s)	ITBR^{-AC}(s)	ITBR^{-PC}(s)	ITBR(s)
repairEx	0.10	0.08	0.08	*0.06*
reviewing	0.33	0.42	0.14	*0.10*
bpic2017 (offer)	0.37	0.42	0.30	*0.30*
receipt	0.17	0.15	0.12	*0.09*
roadtraffic	1.58	2.08	1.18	*1.03*
Billing	2.23	1.91	1.45	*1.36*
bpic2017 (application)	75.7	69.1	64.3	*56.1*
bpic2018	164.8	161.8	158.9	*145.8*
bpic2019	48.6	37.8	43.2	*27.0*

In Table 2, the effectiveness of the implementation is evaluated in order to understand how the improvements in the implementation contribute to the overall efficiency of the approach. Columns in the table represent the execution time of the replay approach when no caching, only post-fix caching, only activity caching and the sum of post-fix caching and activity caching is deployed. In the vast majority of logs, the comparison shows that ITBR provides the best performance.

4.3   Evaluation: Comparison Between Fitness Values

Table 3. Fitness values comparison between the considered approaches on models extracted by the alpha miner and the inductive miner. Here, the first column is the name of the log, from the second to the seventh there is the fitness (whether the algorithms succeed) calculated for the different approaches on models extracted by the inductive miner, from the eight to the tenth there is the fitness calculated by the different considered token-based replay approaches on models extracted by the alpha miner.

	Inductive miner						Alpha miner		
Log	ITBR	CTBR	ABR	BTBR	AFA	REABR	ITBR	CTBR	BTBR
repairEx	1.0	1.0	1.0	1.0	1.0	1.0	0.88	0.88	0.88
reviewing	1.0	1.0	1.0	1.0	1.0	1.0	1.0	1.0	1.0
bpic2017 (offer)	1.0		1.0	1.0	1.0	1.0	0.72		0.72
receipt	1.0	1.0	1.0	1.0	1.0	1.0	0.39	0.39	0.39
roadtraffic	1.0		1.0		1.0	1.0	0.62		0.62
Billing	1.0		1.0	1.0	1.0	1.0	0.69		0.69

In Table 3, a comparison between the fitness values recorded by the ITBR, the CTBR and the ABR is provided, for both the alpha miner and inductive miner models. From then onwards, the biggest logs (bpic2017 (application), bpic2018, bpic2019) are dropped, since a qualitative evaluation is performed. For some real-life logs (bpic2017, roadtraffic, Billing) the CTBR did not succeed in the replay in a reasonable time (an empty space has been reported in the corresponding columns). Alignments have not been evaluated on the models extracted by the alpha miner since it is not assured to have a sound workflow net to start with. The fitness values obtained in Table 3 show that the ITBR, on these logs and the models extracted from them by the inductive miner, is as effective in exploring invisible transitions as the CTBR and the ABR.

4.4   Evaluation: Comparison Between Outputs

A comparison between the output of token-based replay and alignments has been proposed in Table 4. Some popular logs, that are taken into account also for previous evaluations, are being filtered in order to discover a model (using the inductive miner) that is not perfectly fit against the original log. Instead of comparing the fitness values, the comparison is done on the similarity between

Table 4. Comparison between the output of the ITBR and the ABR. First, the name of the log is reported. Then, the number of transitions activated by the two methods is reported, and some aggregations of the similarity measure are provided. Rightmost, the fitness values are reported.

	Transitions		Similarity			Fitness	
Log	ABR	ITBR	Min	Max	Med	ABR	ITBR
repairEx	34858	30459	0.75	1.0	0.94	0.934	0.941
reviewing	9412	8912	0.81	1.0	0.937	0.967	0.974
bpic2017 (offer)	257970	258565	1.0	1.0	1.0	0.995	0.996
receipt	27375	26642	0.42	1.0	0.94	0.839	0.863
roadtraffic	1184482	1023901	0.35	1.0	0.625	0.791	0.816

the set of transitions that are activated in the model during the alignments and the set of transitions that are activated in the model during the token-based replay. The more similar the two sets are, the higher the value of similarity should be. The similarity is calculated as the ratio of the size of the intersection of the two sets and the size of the union of the two sets. This is a simple approach, with some limitations: 1) transitions are counted once during the replay 2) the order in which transitions are activated is not important 3) the number of transitions activated by the alignments is intrinsically higher: while token-based replay could just insert missing tokens and proceed, alignments have to find a path in the model from the initial marking to the final marking, so a higher number of transitions is expected. This comparison, aside fitness values, confirm that the results of the two replay operations, that is a set of transitions activated in the model, are similar. Table 4 provides some further evidence that the two replay techniques are comparable.

4.5   Evaluation: Handling of the Token-Flooding Problem

Table 5. Handling of the token-flooding problem: evaluation between outputs with ($ITBR^{+TFC}$) and without token-flooding cleaning. With the approach enabled, more similar results to alignments are obtained. In the table, the fitness values are reported. Then, in the middle columns, the number of transitions enabled by the methods are inserted. Eventually, the median of the similarity values, as in Sect. 4.4, is reported.

	Fitness			Transitions			Similarity	
Log	ABR	ITBR	ITBR^{+TFC}	ABR	ITBR	ITBR^{+TFC}	ITBR	ITBR^{+TFC}
repairEx	0.934	0.941	0.934	34858	30459	30459	0.94	0.94
reviewing	0.967	0.974	0.967	9412	8912	8912	0.937	0.937
bpic2017 (offer)	0.995	0.996	0.995	257970	258565	259597	1.0	1.0
receipt	0.839	0.863	0.862	27375	26642	27508	0.94	0.94
roadtraffic	0.791	0.816	0.791	1184482	1023901	1184039	0.625	0.625

In Table 5, the importance of handling the token flooding problem is illustrated on several logs. The models against which the technique is evaluated are the same obtained in Sect. 4.4. For both the fitness values (albeit the underlying concepts/fitness formulas are different) and the number of transitions activated in the model, we are getting a more similar (higher) number, since the activation of unwanted parts of the process model is avoided. For the median of similarity between the outputs, we obtain equal numbers between the ITBR and the ITBR^{+TFC} approach; this means that the token flooding procedure acts only on the most problematic traces of the log according to the model.

```
1  from pm4py.objects.log.importer.xes import factory as xes_importer
2  from pm4py.algo.discovery.alpha import factory as alpha_miner
3  from pm4py.algo.conformance.token_replay import factory as tr_factory
4  log = xes_importer.apply("C:\\running-example.xes")
5  net, im, fm = alpha_miner.apply(log)
6  aligned_traces = tr_factory.apply(log, net, im, fm)
```

Fig. 5. PM4Py code to load a log, apply the alpha miner and visualize a Petri net.

5 Tool Support

The contribution described in this paper has been implemented in the Python library PM4Py. The tool can be easily installed in the Python 3.7 environment following the documentation reported on the website. The application of token-based replay is performed on an event log and an accepting Petri net. Example code to import a XES file, apply the alpha miner and then the token-based replay is presented in Fig. 5.

In the tool, we provide also some advanced diagnostics, in order to be able to answer to the following questions:

1. If a given transition is executed in an unfit way, what is the effect on the throughput time?
2. If a given transition is executed in an unfit way, why does this happen?
3. If a given activity that is not contained in the process model is executed, what is the effect on the throughput time?
4. If a given activity that is not contained in the process model is executed, why does this happen?

For questions 1) and 3), the *throughput time* diagnostic introduced in Sect. 5.1 can be used. For questions 2) and 4), the *root cause analysis* diagnostic introduced in Sect. 5.2 can provide the corresponding answers.

The documentation about the usage of the token-based replay[5] and of the diagnostics[6] is available on the website.

---

[5] http://pm4py.pads.rwth-aachen.de/documentation/conformance-checking/token-based-replayer/.

[6] http://pm4py.pads.rwth-aachen.de/documentation/conformance-checking/token-based-replayer/token-based-replay-diagnostics/.

5.1 Advanced Diagnostics: Throughput Time Analysis

The comparison between the throughput time in non-fitting cases and fitting cases permits to understand, for each kind of deviations, whether it is important or not important for the throughput time. To evaluate this, the "Receipt phase of an environmental permit application process" log is taken. After some filtering operations, the model represented in Fig. 1 is obtained. Several activities that are in the log are missing according to the model, while some transitions have fitness issues. After performing the token-based replay enabling the local information retrieval, and applying the *duration_diagnostics.diagnose_from_trans_fitness* function to the log and the transitions fitness object, it can be seen that the transition *T06 Determine necessity of stop advice* is executed in an unfit way in 521 cases. For the cases where this transition is enabled according to the model the median throughput time is around 20 min, while in the cases where this transition is executed in an unfit way the median throughput time is 1.2 days. So, the throughput time of unfit cases is 146 times higher in median than the throughput time of fit cases. Activities of the log that are not in the model are likely to make the throughput time of the process higher since they are executed rarely. In our implementation, applying the *duration_diagnostics.diagnose_from_notexisting_activities* method, the median execution time of cases containing these activities can be retrieved and compared with the median execution time of cases that do not contain them (that is 20 minutes). Taking into account the activity *T12 Check document X request unlicensed*, it is contained in 44 cases, which median throughput time is 6.9 days (505 times higher than the standard).

5.2 Advanced Diagnostics: Root Cause Analysis

Root cause analysis is a type of diagnostic, that is obtained on top of the token-based replay results, that permits to understand the reasons why a deviation happened. This is done using the ideas of the framework described in [12]:

- Log attributes (at the case and the event level) are transformed into numeric features (for string attributes, one-hot encoding is applied); for each case, a vector of features is obtained.
- A class (e.g. for (2), 0 for fit traces, 1 for unfit traces; for (4), 0 for traces not containing the activity, 1 for traces containing it) is assigned to each case.
- A machine learning algorithm is applied to learn a representation of the data.

By transforming the log into a matrix of numeric features, interoperability is kept across a wide set of machine learning classification algorithms (e.g. decision trees, random forests, deep learning methods). Within our implementation, decision trees are used to get a description of the differences between the two classes. The decision tree in our approach was trained on the entire dataset, since the goal is to obtain some discrimination rules between the two classes.

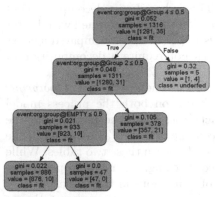

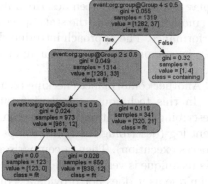

a) Decision tree extracted comparing fit and unfit cases for transition *T02 Check Confirmation of receipt*.

b) Decision tree extracted comparing cases containing and not containing activity *T03 Adjust confirmation of receipt* that is not in the model.

Fig. 6. Root Cause Analysis performed on the log "Receipt phase of an environmental permit application process" and a model extracted using inductive miner on a filtered version of the log. In the represented decision trees, two different kind of deviations, a) an activity that is in the model but is executed in an unfit way, and b) an activity is executed that is not in the model, have been analyzed.

This framework permits to answer to the following questions:

1. If a given transition is executed in an unfit way, why does this happen?
2. If a given activity that is not contained in the process model is executed, which is the effect on the throughput time?

To evaluate this, the "Receipt phase of an environmental permit application process" log and the model represented in Fig. 1 are taken. In the following examples, the decision tree has been built using only the *org:group* attribute. Applying the *root_cause_analysis.diagnose_from_notexisting_activities* method, for transition *T02 Check Confirmation of receipt* the decision tree shown in Fig. 6(a) is obtained, that permits to understand the following information: (1) Group 4 triggers an unfit execution according to the model. (2) Group 2 triggers sometimes an unfit execution according to the model. Applying the *duration_diagnostics.diagnose_from_notexisting_activities* method, for activity *T03 Adjust confirmation of receipt* the decision tree shown in Fig. 6(b) is obtained, that permits to understand that Group 4 and 2 trigger the activity.

6 Related Work

Token-based replay has been introduced as a conformance checking technique in [27]. The approach has also been used internally in some process discovery algorithms such as the genetic miner [19] to evaluate the quality of the candidates. Recently, a flexible online replay technique, that provides token-based replay as option, has been described in [10]. This is based on a decomposition of the model, in such way the state space exploration can be performed with better performance. The approach introduced in this paper has been compared against [27]; in comparison to [10], our approach does not require a decomposition of the model.

Another conformance checking technique for Petri nets is the one of *footprints* [4]. In this technique, a footprint table is found on both the process model (describing the relationships between the activities as in the model) and the event log (describing the relationships between the activities as recorded in the process execution). Then, a comparison is done between these two tables. While this technique is very scalable for conformance checking, it is not a proper replay technique as it does not provide a sequence of transitions in the model.

Currently, the standard replay technique on Petri nets is the computation of *alignments with optimal cost* [4,11]. In the assessment, we have compared against the approach described in [7], showing that our token-based replay provides better performance than such technique.

Other techniques are based on decomposing the model [2,22], in order to perform a multiple number of smaller alignments. The recomposition approach described in [15] is able to provide the optimal cost of an alignment between the model and the process execution under some assumptions. The technique usually leads to shorter execution times. However, token-based replay is often still faster (as shown in the assessment).

Approaches to approximate the conformance checking results are described in [8,29]; these might not produce the optimal cost of an alignment but produce generally a good approximation of the alignment or of its cost. In comparison, our approach is able to produce a proper path in the model when the execution is fit (see the assessment).

In [3,28], map-reduce approaches have been applied to parallelize the computation of the alignments. Online conformance checking techniques [31] iteratively update the alignment to include new events; in doing so, for efficiency reasons, the number of states stored and visited might be reduced, hence optimality of the alignments is not granted. The improved token-based replay approach introduced in the paper is an offline technique. At the moment, we don't provide any scalable map-reduce architecture.

Other replay techniques have focused on different types of process models. In [17], a process tree discovered using inductive miner is converted into a deterministic finite automaton for fast fitness checking. In [24], the goal is to perform alignments on automatons. This shows some advantages in models without concurrency, but suffer from scalability issues in models with concurrency. In [23], a decomposition of a Petri net model into S-components is performed in order to get a collection of automatons, against which alignments are performed. In [21], an efficient replay technique for BPMN models is proposed.

Table 6. A description of the replay techniques presented in Sect. 6. The third column is the target model. The fourth column describes the super-class of the replay technique (Ali = alignments, TR = token-based replay, DFA = DFA semantics, FP = footprints). The fifth column describes whether the technique is an online technique. The sixth column describes whether the output is optimal (see the bottom of Sect. 6).

Refs.	Description	Model	Appr.	Online	Opt.
[9, 19, 27]	Token-based replay approaches	Petri nets	TR	No	No
[10]	Flexible conformance checking approach, based on a decomposition	Petri nets	TR/Ali	Yes	No
[7]	Alignments with optimal cost	Petri nets	Ali	No	Yes
[15]	Replay technique based on decomposing the model, performing alignments and recomposing the result	Petri nets	Ali	No	Yes/No
[8, 29]	Different replay techniques based on alignments approximation	Petri nets	Ali	No	No
[3, 28]	Distributed alignments computation	Petri nets	Ali	No	Yes
[31]	Online conformance checking	Petri nets	Ali	Yes	No
[24]	Alignments on top of automatons	DFA	Ali	No	Yes
[17]	Technique to verify the fitness of traces on top of process trees through conversion to a finite automaton	DFA	DFA	No	No
[23]	Alignment technique that exploits a decomposition of the original Petri net model in S-components, converts them in finite automatons, and apply alignment on the single components	DFA	Ali	No	No
[21]	Replay technique on top of BPMN	BPMN	TR	No	No
[4]	Footprints comparison	Any	FP	No	No

Table 6 summarizes the approaches discussed in this section. The optimality concept is defined only for the techniques producing alignments (see [11]). Since token-based replay techniques are based on heuristics for invisible/duplicate transitions, and the footprints technique is a matrix comparison, they have been considered as non-optimal.

7   Conclusion

In this paper, an improved token-based replay approach for Petri nets has been proposed. The technique exploits a preprocessing step that leads to a better handling of invisible transitions. Moreover, the intermediate storage techniques have been improved to achieve a lower execution time.

Token-based replay approaches already outperformed alignment-based approaches for Petri nets with visible transitions. The proposed token-based replay approach is faster than alignment-based approaches for Petri nets also for models with invisible transitions.

Next to an increase in speed, the problem of token flooding is addressed by "freezing" superfluous tokens (see Sect. 3.3). In this way, the replay does not lead to markings with many more tokens than what would be possible according to the model, avoiding the activation of unwanted parts of the process model and leading to lower values of fitness for problematic parts of the model. Moreover, we showed that we are able to diagnose the effects of deviations on the case throughput time, and we are able to perform root cause analysis.

The approach has some limitations. First, we do not propose any termination or fitness guarantees. Also, performance is in some cases worse than advanced replay techniques as automaton-based alignments (as AFA). However, the improved token-based replay has a clear performance lead on the biggest logs and models that have been considered (BPI Challenge 2017, 2018 and 2019).

We hope that this will trigger a revival of token-based replay, a technique that seemed abandoned in recent years. Especially when dealing with large logs, complex models, and real-time applications, the flexible tradeoff between quality and speed provided by our implementation is beneficial.

Acknowledgements. We thank the Alexander von Humboldt (AvH) Stiftung for supporting our research.

References

1. van der Aalst, W.M.P.: Structural characterizations of sound workflow nets. Comput. Sci. Rep. **96**(23), 18–22 (1996)
2. van der Aalst, W.M.P.: Decomposing process mining problems using passages. In: Haddad, S., Pomello, L. (eds.) PETRI NETS 2012. LNCS, vol. 7347, pp. 72–91. Springer, Heidelberg (2012). https://doi.org/10.1007/978-3-642-31131-4_5
3. van der Aalst, W.M.P.: Distributed process discovery and conformance checking. In: de Lara, J., Zisman, A. (eds.) FASE 2012. LNCS, vol. 7212, pp. 1–25. Springer, Heidelberg (2012). https://doi.org/10.1007/978-3-642-28872-2_1
4. van der Aalst, W.M.P.: Data science in action. Process Mining, pp. 3–23. Springer, Heidelberg (2016). https://doi.org/10.1007/978-3-662-49851-4_1
5. van der Aalst, W.M.P., Adriansyah, A., van Dongen, B.: Replaying history on process models for conformance checking and performance analysis. Wiley Interdisc. Rev. Data Min. Knowl. Discov. **2**(2), 182–192 (2012)
6. Adriansyah, A.: Aligning observed and modeled behavior. Ph.D. thesis, Department of Mathematics and Computer Science (2014)
7. Adriansyah, A., Sidorova, N., van Dongen, B.: Cost-based fitness in conformance checking. In: 2011 11th International Conference on Application of Concurrency to System Design (ACSD), pp. 57–66. IEEE (2011)
8. Bauer, M., van der Aa, H., Weidlich, M.: Estimating process conformance by trace sampling and result approximation. In: Hildebrandt, T., van Dongen, B.F., Röglinger, M., Mendling, J. (eds.) BPM 2019. LNCS, vol. 11675, pp. 179–197. Springer, Cham (2019). https://doi.org/10.1007/978-3-030-26619-6_13
9. Berti, A., van der Aalst, W.M.P.: Reviving token-based replay: increasing speed while improving diagnostics. In: Algorithms & Theories for the Analysis of Event Data (ATAED 2019) (CEUR 2371), pp. 87–103 (2019)
10. vanden Broucke, S.K.L.M., Munoz-Gama, J., Carmona, J., Baesens, B., Vanthienen, J.: Event-based real-time decomposed conformance analysis. In: Meersman, R., et al. (eds.) OTM 2014. LNCS, vol. 8841, pp. 345–363. Springer, Heidelberg (2014). https://doi.org/10.1007/978-3-662-45563-0_20

11. Carmona, J., Dongen, B., Solti, A., Weidlich, M.: Conformance Checking: Relating Processes and Models. Springer, Heidelberg (2018). https://doi.org/10.1007/978-3-319-99414-7
12. de Leoni, M., van der Aalst, W.M.P., Dees, M.: A general framework for correlating business process characteristics. In: Sadiq, S., Soffer, P., Völzer, H. (eds.) BPM 2014. LNCS, vol. 8659, pp. 250–266. Springer, Cham (2014). https://doi.org/10.1007/978-3-319-10172-9_16
13. Esparza, J.: Reduction and synthesis of live and bounded free choice petri nets. Inf. Comput. **114**(1), 50–87 (1994)
14. Kerremans, M.: Gartner Market Guide for Process Mining, Research Note G00353970 (2018). www.gartner.com
15. Lee, W.L.J., Verbeek, H., Munoz-Gama, J., van der Aalst, W.M.P., Sepúlveda, M.: Recomposing conformance: closing the circle on decomposed alignment-based conformance checking in process mining. Inf. Sci. **466**, 55–91 (2018)
16. Leemans, S.J.J., Fahland, D., van der Aalst, W.M.P.: Discovering block-structured process models from event logs - a constructive approach. In: Colom, J.-M., Desel, J. (eds.) PETRI NETS 2013. LNCS, vol. 7927, pp. 311–329. Springer, Heidelberg (2013). https://doi.org/10.1007/978-3-642-38697-8_17
17. Leemans, S.J., Fahland, D., van der Aalst, W.M.P.: Scalable process discovery and conformance checking. Softw. Syst. Model. **17**(2), 599–631 (2018)
18. Martínez, J.: A simple and fast algorithm to obtain all invariants of a generalised Petri net. In: Girault, C., Reisig, W. (eds.) Application and Theory of Petri nets. INFORMATIK, vol. 52, pp. 301–310. Springer, Heidelberg (1982). https://doi.org/10.1007/978-3-642-68353-4_47
19. de Medeiros, A.K.A., Weijters, A.J., van der Aalst, W.M.P.: Genetic process mining: an experimental evaluation. Data Min. Knowl. Disc. **14**(2), 245–304 (2007)
20. Miyamoto, T., Kumagai, S.: Calculating place capacity for Petri nets using unfoldings. In: International Conference on Application of Concurrency to System Design (ACSD) 1998, pp. 143–151. IEEE (1998)
21. Molka, T., Redlich, D., Drobek, M., Caetano, A., Zeng, X.J., Gilani, W.: Conformance checking for BPMN-based process models. In: Proceedings of the 29th Annual ACM Symposium on Applied Computing, pp. 1406–1413 (2014)
22. Munoz-Gama, J., Carmona, J., van der Aalst, W.M.P.: Single-entry single-exit decomposed conformance checking. Inf. Syst. **46**, 102–122 (2014)
23. Reißner, D., Armas-Cervantes, A., Conforti, R., Dumas, M., Fahland, D., La Rosa, M.: Scalable alignment of process models and event logs: an approach based on automata and s-components. arXiv preprint arXiv:1910.09767 (2019)
24. Reißner, D., Conforti, R., Dumas, M., La Rosa, M., Armas-Cervantes, A.: Scalable conformance checking of business processes. OTM 2017. LNCS, vol. 10573, pp. 607–627. Springer, Cham (2017). https://doi.org/10.1007/978-3-319-69462-7_38
25. Rogge-Solti, A., Senderovich, A., Weidlich, M., Mendling, J., Gal, A.: In log and model we trust? A generalized conformance checking framework. In: La Rosa, M., Loos, P., Pastor, O. (eds.) BPM 2016. LNCS, vol. 9850, pp. 179–196. Springer, Cham (2016). https://doi.org/10.1007/978-3-319-45348-4_11
26. Rozinat, A., van der Aalst, W.M.P.: Conformance testing: measuring the fit and appropriateness of event logs and process models. In: Bussler, C.J., Haller, A. (eds.) BPM 2005. LNCS, vol. 3812, pp. 163–176. Springer, Heidelberg (2006). https://doi.org/10.1007/11678564_15
27. Rozinat, A., van der Aalst, W.M.P.: Conformance checking of processes based on monitoring real behavior. Inf. Syst. **33**(1), 64–95 (2008)

28. Shugurov, I., Mitsyuk, A.: Applying MapReduce to conformance checking. Proc. Inst. Syst. Program. RAS **28**, 103–122 (2016). https://doi.org/10.15514/ISPRAS-2016-28(3)-7
29. Taymouri, F., Carmona, J.: An evolutionary technique to approximate multiple optimal alignments. In: Weske, M., Montali, M., Weber, I., vom Brocke, J. (eds.) BPM 2018. LNCS, vol. 11080, pp. 215–232. Springer, Cham (2018). https://doi.org/10.1007/978-3-319-98648-7_13
30. Weijters, A., van der Aalst, W.M.P., De Medeiros, A.A.: Process mining with the heuristics miner-algorithm, pp. 1–34. Technical report WP 166, Technische Universiteit Eindhoven (2006)
31. van Zelst, S.J., Bolt, A., Hassani, M., van Dongen, B.F., van der Aalst, W.M.P.: Online conformance checking: relating event streams to process models using prefix-alignments. Int. J. Data Sci. Anal. **8**(3), 269–284 (2019). https://doi.org/10.1007/s41060-017-0078-6

Extensible Structural Analysis of Petri Net Product Lines

Elena Gómez-Martínez[(✉)][iD], Juan de Lara[iD], and Esther Guerra[iD]

Modelling and Software Engineering Research Group,
Universidad Autónoma de Madrid, C/ Francisco y Valiente, 11, 28049 Madrid, Spain
{MariaElena.Gomez,Juan.DeLara,Esther.Guerra}@uam.es
http://miso.es

Abstract. Petri nets are a popular formalism to represent concurrent systems. However, their standard form does not offer variability support to model and effectively analyse large sets of variants of a given system. For this purpose, we propose a notion of *product line* of Petri nets to represent a set of similar concurrent systems. The formalization enriches Petri nets with a feature model characterizing the variability of the systems. Moreover, places, transitions and arcs can define presence conditions that determine the subset of system variants they belong to.

To enable an efficient analysis of the set of all net variants, we have lifted several structural analysis methods for Petri nets, to the product line level. Currently, we support the lifted checking of the marked graph, state-machine, and (extended) free-choice properties, which avoids their analysis on each particular net of the product line in isolation.

We demonstrate the feasibility of our proposal using examples in the domain of flexible assembly lines, and introduce an extensible tool infrastructure. The tool is based on Eclipse and FeatureIDE, and permits adding new analysis methods externally. Moreover, we present an evaluation that shows the efficiency gains of our method with respect to an enumerative approach that analyses the properties on every net within the product line separately.

Keywords: Petri nets · Structural analysis · Product lines · Model-driven engineering

1 Introduction

Petri nets are a popular formalism to model concurrent systems [21]. They are widely used due to their rich body of theoretical results enabling analysis, and the plethora of existing supporting tools[1]. However, some scenarios require modelling (possibly a large set of) variants of similar systems. Some examples reported in the literature include the design of the variants of controllers for cyber-physical systems [20], modelling all possible variants of flexible assembly lines [24], or

---

[1] See for example https://www.informatik.uni-hamburg.de/TGI/PetriNets/tools/.

© Springer-Verlag GmbH Germany, part of Springer Nature 2021
M. Koutny et al. (Eds.): ToPNoC XV, LNCS 12530, pp. 27–49, 2021.
https://doi.org/10.1007/978-3-662-63079-2_2

building families of workflow process models [28]. In these cases, the designer needs to build many variations of a base model. However, if there are many variants, then building, maintaining and analysing this large set of variants becomes challenging.

To facilitate the management of large sets of net variants, we combine Petri nets with software product lines (SPLs) [25,27] to define a notion of Petri net product line (PNPL). This allows modelling the variability space using a feature model, and automatically producing specific Petri nets from given feature configurations [15].

As the main contribution of this paper, we propose lifting some structural analysis techniques of Petri nets to the product line level. This means that we do not need to analyse each Petri net that can be produced from a PNPL separately, but our analysis techniques work on the whole set of Petri nets directly. In this paper, we explain how to lift the analysis of the marked graph, state-machine, and (extended) free-choice [11] properties to PNPLs, but other structural analysis techniques like (extended) asymmetric choice [2] or equal conflict nets [32] can be lifted in a similar way. In the above-mentioned scenarios, these structural analysis techniques can be used to assess soundness of workflow nets [1] by analysing if some/all nets are free choice; to check whether synchronization can interfere with conflicts in a flexible assembly line by analysing if some/all nets are free choice; or checking whether any variant of a controller design can lead to conflicts by checking if all variants are marked graphs.

As a second contribution, we present extensible prototype tool support to model and analyse PNPLs. Our tool is based on Eclipse, and has an extension point to enable contributing further analysis techniques externally. Moreover, we use our tool to evaluate the efficiency of our lifted analysis techniques, which show good improvement compared to enumerating and analysing each Petri net within the product line.

This paper extends our work in [12] by a more comprehensive formalization of the lifting process (Sect. 3.1), the lifted analysis of two additional properties (free-choice and extended free-choice), improved tool support and an expanded evaluation.

In the following, Sect. 2 introduces PNPLs; Sect. 3 proposes lifting the analysis of structural properties to PNPLs and lifts the analysis of the marked graph and (extended) free-choice properties; Sect. 4 presents tool support; Sect. 5 evaluates the efficiency of our lifted analysis; Sect. 6 compares with related research; and Sect. 7 concludes. The Appendix details the lifting of the state-machine property.

2   Petri Net Product Lines

This section defines PNPLs, and how to derive concrete Petri nets via feature configurations. We consider a simple notion of Petri net, as given in Definition 1, but the approach can be easily adapted to other more complex versions.

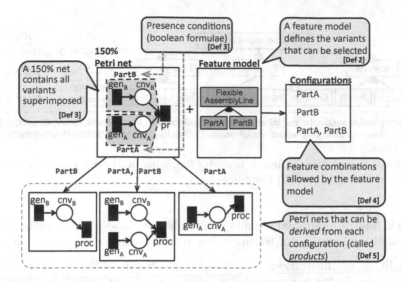

Fig. 1. Ingredients of a PNPL.

Definition 1 (Petri net). *A Petri net is a tuple* $PN = (P, T, A)$, *where* P *and* T *are disjoint sets of places and transitions, and* $A \subseteq (P \times T) \cup (T \times P)$ *is the set of arcs connecting either places to transitions or vice versa.*

Given an arc $a \in A$, we use a_0 to refer to its source, and a_1 to refer to its target.

We define a notion of PNPL to support the definition of net variants. Figure 1 shows the concepts that it involves. Firstly, the variability space is represented as a set of features in a feature model. Then, the main idea is to superimpose all net variants within a single net – called 150% Petri net[2] – and annotate its elements with presence conditions (logic formulae over the features in the feature model). Users can retrieve a particular net variant by selecting a subset of the available features. Such a choice is called a configuration. Then, the selected features are substituted by *true* in the presence conditions, and the unselected ones by *false*. This makes each presence condition to evaluate either to true or false. The elements whose presence condition evaluates to false are eliminated from the 150% net, and the remaining elements form the selected net variant. In the following, we define each component of the approach in detail.

PNPLs build on the notion of a feature model that defines the variability space of possible configurations.

Definition 2 (Feature model). *A feature model* $FM = (F, \Psi)$ *consists of a set of propositional variables* $F = \{f_1, ..., f_n\}$ *called* features, *and a propositional formula* Ψ *over the variables in* F.

---

[2] The term 150% model is standard in software product lines. It refers to the fact that a single model contains many variants superimposed.

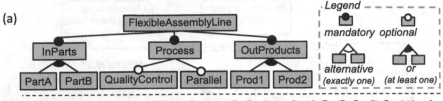

(b) FM=⟨{FlexibleAssemblyLine, InParts, Process, OutProducts, PartA, PartB, QualityControl, ...},
FlexibleAssemblyLine∧InParts∧Process∧OutProducts∧(PartA∨PartB)∧(Prod1∨Prod2)⟩

Fig. 2. Feature model for the flexible assembly line using (a) the diagrammatic notation of feature models [15], and (b) Definition 2.

Remark. The propositional formula Ψ in the feature model is used to determine the allowed combinations of feature values (those making the formula true).

Example. As an illustration, we will be using a family of Petri nets describing the behaviour of a flexible assembly line, that is, a production system that can be quickly reconfigured in different set-ups to produce a variety of goods or adapt to customer demands [24]. Here, the problem is to model all such possible configurations in a compact way, and analyse properties of all configurations efficiently. Figure 2(a) shows the feature model using a diagrammatic notation [15], and Fig. 2(b) using Definition 2. Our assembly line can be configured to accept one or two kinds of input parts (PartA, PartB), can optionally have a quality control process (QualityControl) and a parallel conveyor (Parallel), and can produce one or two kinds of products (Prod1, Prod2).

A PNPL is a Petri net whose elements can be annotated with boolean formulae, having as variables the features of the feature model.

Definition 3 (Petri net product line). *A PNPL $PNL = (FM, PN, \Phi)$ is made of a feature model FM, a Petri net PN (called the 150% Petri net), and a mapping Φ which consists of pairs $\langle x, \Phi_x \rangle$ mapping an element $x \in P \cup T \cup A$ to a propositional formula Φ_x (called the presence condition (PC) of x) over the features in FM.*

PNL is well-formed if $\forall a \in A: (\Phi_a \Rightarrow \Phi_{a_0}) \wedge (\Phi_a \Rightarrow \Phi_{a_1})$ is true.

As noticed, we use an annotative approach to facilitate the analysis. The approach relies on the definition of a 150% Petri net that contains all variants of the PNPL, and the assignment of PCs to its elements. Then, a particular Petri net can be obtained by removing the elements whose PC evaluates to false given a choice (a *configuration*) of feature values. This kind of variability which starts from a maximal description of a set of systems (the 150% Petri net) and deletes elements upon certain conditions is called *negative* [10]. Instead, other approaches to SPLs use *positive* variability, i.e., they start from a minimal description of the systems to which new elements are added depending on the selected features [29]. Our method can also be applied to positive variability approaches as long as they permit deriving a 150% Petri net.

In Definition 3, the well-formedness condition requires the PC of an arc to be stronger than the PC of its source and target elements. This ensures that, if the arc is present in a product Petri net (i.e., a Petri net derived from a configuration by deleting from the 150% net the elements whose PC is false), its source and target elements will be present as well. Definitions 4 and 5 will provide the formal notions of configuration and Petri net derivation.

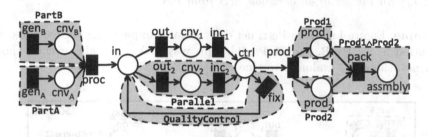

Fig. 3. 150% Petri net with PC annotations, modelling a flexible assembly line.

Example. Figures 2 and 3 show the feature model and the 150% net composing the PNPL of the flexible assembly line. The 150% Petri net in Fig. 3 uses dashed regions as a shortcut to assign the same PC to all the elements in the region. For example, formula PartB in the top-left corner is attached to transition gen_B, to place cnv_B, and to the arcs from/to place cnv_B. If an element does not show an attached PC, then we assume that its PC is true.

The way to obtain a specific *product* Petri net from a PNPL is by selecting a subset of the features in its feature model. This selection is called a *feature configuration*. In the following definition, we use $\Psi[X/true, Y/false]$ to denote the substitution of all variables in X by *true*, and all variables in Y by *false*, in formula Ψ.

Definition 4 (Feature configuration). *A valid feature configuration $\rho \subseteq F$ of a PNPL PNL with feature model $FM = (F, \Psi)$ is a subset of its features satisfying Ψ, i.e., $\Psi[\rho/true, F \setminus \rho/false]$ evaluates to true when each $f \in \rho$ is substituted by true, and each $f \in F \setminus \rho$ is substituted by false. We use $P(FM) = \{\rho_i\}$ for the set of all valid feature configurations of PNL.*

To improve readability, in the remaining of the paper, feature configurations omit features that are mandatory in any configuration.

Example. Figure 2 admits 36 feature configurations. In all of them, PartA or PartB (inclusive) need to be selected, and similarly, Prod1 or Prod2 need to be selected as well. For instance, some valid configurations are $\rho_0 = \{\text{PartA}, \text{Prod1}\}$, $\rho_1 = \{\text{PartA}, \text{PartB}, \text{Prod1}\}$, and $\rho_2 = \{\text{PartB}, \text{Parallel}, \text{Prod1}\}$. As mentioned above, these configurations would also include features FlexibleAssemblyLine, InParts, Process and OutProducts, but we do not show them as they are mandatory.

Given a feature configuration, we obtain the corresponding product Petri net by removing from the 150% Petri net any element whose PC is false.

Definition 5 (Petri net derivation). *Given a PNPL PNL = (FM, PN = (P, T, A), Φ) and a configuration ρ ∈ P(FM), we derive the net $PN_\rho = (P_\rho, T_\rho, A_\rho)$ building each set $X_\rho \subseteq X$ (for X = {P, T, A}) as {x ∈ X | Φ_x[ρ/true, F \ ρ/false] = true}. We use Prod(PNL) = {PN_ρ | ρ ∈ P(FM)} for the set of all derivable nets from PNL.*

Example. Figure 4 shows a Petri net derivation example using the feature configuration ρ_2 = {PartB, Parallel, Prod1}. This way, PN_{ρ_2} contains exactly those elements whose PC evaluates to true.

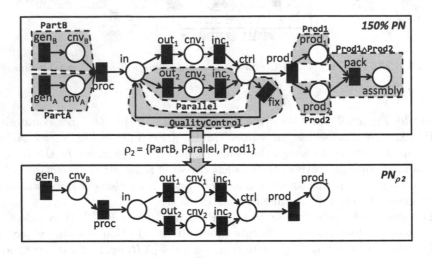

Fig. 4. Petri net derivation example.

To analyse a property in every Petri net that can be derived from a PNPL, a naive method would derive and analyse each product Petri net one by one. However, this can be time-consuming since the number of derivable Petri nets can be exponential on the number of features in the worst case. Hence, the next section proposes a method to lift the analysis of structural properties to the product line level.

3   Structural Analysis of Petri Net Product Lines

This paper is focused on the efficient analysis of structural properties of the set of nets that can be derived from a PNPL. Structural properties depend only on the net topology and are independent of the initial marking [21]. These properties include connectedness, state-machine, marked graph, and (extended) free-choice, among others.

In the following, we first introduce the general scheme and required concepts for the lifted analysis of structural properties (Sect. 3.1). Then, we lift the analysis for the marked graph (Sect. 3.2), free-choice (Sect. 3.3) and extended free-choice properties (Sect. 3.4). The appendix contains the lifting of the state-machine property.

3.1 Lifting the Analysis of Structural Properties

Structural properties look at connectivity patterns of a given Petri net to assert the occurrence of some particular structure. These properties are frequently formulated using first-order logic and auxiliary functions, such as the pre- and post-sets of each place and transition in the net. Figure 5(a) illustrates the formalization of a structural property using a formula P, which is checked on a Petri net PN. We write $PN \models P$ to indicate that property P holds on the Petri net PN.

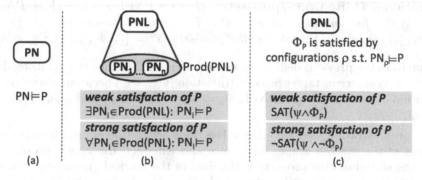

Fig. 5. (a) Checking a structural property P on a Petri net PN. (b) Checking a structural property P on a PNPL PNL using an enumerative approach. (c) Checking a structural property P on a PNPL PNL using a lifted approach.

To check a structural property in the set of nets of a PNPL, we can separately check the property in each derivable net PN_i, as Fig. 5(b) shows. Because we now look at a set of nets, instead of at individual ones, we can distinguish between *weak* and *strong* property satisfaction. Weak satisfaction requires that some product Petri net of the PNPL satisfies the property P (e.g., the marked graph property, cf. Definition 7), while strong satisfaction requires that all product Petri nets satisfy P. The problem with this solution is that checking P on each product net might be too costly as there may be an exponential number of them.

Instead, we propose the solution outlined in Fig. 5(c) to improve the efficiency of the analysis of structural properties for a PNPL. In this solution, we first lift the property P to the product line level. For this purpose, we encode P as a formula Φ_P which takes into account the PCs of the elements in the 150% Petri net, and is satisfied by those configurations ρ such that $PN_\rho \models P$. Then, we recast the checking of weak/strong property satisfaction as a constraint satisfaction problem. Specifically, if $SAT(\Psi \wedge \Phi_P)$ (with SAT a predicate that holds if the formula is satisfiable, and Ψ the formula of the feature model), then there is some valid configuration which produces a Petri net that satisfies the property P. We can use a constraint solver to obtain a feature configuration that satisfies the formula $\Psi \wedge \Phi_P$. If such a configuration exists, then we have weak property satisfaction.

Conversely, the formula $\neg\Phi_P$ is satisfied by those configurations that produce Petri nets where P does not hold. This way, we have strong satisfiability if $SAT(\Psi \wedge \neg\Phi_P)$ does not hold. This means that no valid configuration (satisfying Ψ) produces a Petri net that does not satisfy P (where $\neg\Phi_P$ holds).

The structural properties that we consider in this paper – state-machine, marked graph, (extended) free-choice – make use of the pre- and post-sets of each place p and each transition t (written $^\bullet p$, $p^\bullet$, $^\bullet t$ and $t^\bullet$ respectively). Hence, we need to incorporate the PCs within those sets, as Definition 6 shows.

Definition 6 (Lifted pre-/post-sets). *Given a PNPL $PNL = (FM, PN = (P, T, A), \Phi)$, for any element $x \in P \cup T$, the lifted pre-set of x is $^\circ x = \{(y, \Phi_{(y,x)}) \mid (y, x) \in A\}$, while its lifted post-set is $x^\circ = \{(y, \Phi_{(x,y)}) \mid (x, y) \in A\}$.*

Remark. In the previous definition, we can use the PC of the arc (Φ_a) instead of the PC of its source or target place or transition (Φ_{a_0}, Φ_{a_1}) because, according to Definition 3, in a well-formed PNPL, $\Phi_a \Rightarrow \Phi_{a_0} \wedge \Phi_a \Rightarrow \Phi_{a_1}$, and so, $\Phi_a \wedge \Phi_{a_0} = \Phi_a = \Phi_a \wedge \Phi_{a_1}$.

As an illustration, the following subsections apply this approach to lift the analysis of the marked graph, free-choice and extended free-choice properties. Since the state-machine property is the dual of the marked graph property, we show it in the Appendix. Other structural properties like asymmetric choice can be lifted in a similar way.

3.2   Lifted Analysis of the Marked Graph Property

Firstly, we provide the definition of the marked graph (MG) property. In a MG Petri net, each place has exactly one input transition and one output transition, whereas each transition may have multiple input and output places. Therefore, a MG allows concurrent and synchronization structures with no conflict.

Definition 7 (Marked graph, from [21]). *A Petri net $PN = (P, T, A)$ is a marked graph, written $PN \models MG$, if $\forall p \in P : |^\bullet p| = |p^\bullet| = 1$.*

We lift this definition of MG to the product line level. Therefore, a PNPL strongly (weakly) satisfies the MG property if all (some of) its derivable nets are MGs.

Definition 8 (Strong and weak MG product line). *A Petri net product line PNL is a strong marked graph iif* $\forall PN_\rho \in Prod(PNL) :\ PN_\rho \models MG$. *PNL is a weak marked graph iif* $\exists PN_\rho \in Prod(PNL) :\ PN_\rho \models MG$.

If we can derive from the product line PNL a net that is not a MG, then PNL is not a strong MG product line. In particular, given a feature configuration ρ, a Petri net derivation PN_ρ is not a MG if it has a place p with more than one input transition, more than one output transition, no input transitions, or no output transitions. Therefore, for a PNPL to be a strong MG, we require that the size of the lifted pre-set $°p = \{(t_0, \Phi_{(t_0,p)}), ..., (t_n, \Phi_{(t_n,p)})\}$ and the lifted post-set $p° = \{(t_0, \Phi_{(p,t_0)}), ..., (t_n, \Phi_{(p,t_n)})\}$ of every place p to be one for every possible configuration. For the case of the pre-set, this is the case if the following formula is true:

$$
\begin{aligned}
\Phi_{°p} \triangleq\ &false \\
&\vee\ (\Phi_{(t_0,p)} \wedge \neg\Phi_{(t_1,p)} \wedge ... \wedge \neg\Phi_{(t_n,p)}) \\
&\vee\ (\neg\Phi_{(t_0,p)} \wedge \Phi_{(t_1,p)} \wedge ... \wedge \neg\Phi_{(t_n,p)}) \\
&\vee\ ...\ (\neg\Phi_{(t_0,p)} \wedge \neg\Phi_{(t_1,p)} \wedge ... \wedge \Phi_{(t_n,p)})
\end{aligned}
\tag{1}
$$

The formula is made of a disjunction of conjunctions, where only one term in each conjunction can be true. This ensures that, regardless of the configuration, the pre-set of the place will have size one. The disjunctions start with $false$, so that $\Phi_{°p}$ is false when $°p$ is empty. The terms $\Phi_{(t_i,p)}$ are the PCs in the lifted pre-set of p ($°p$). The formula that ensures that the size of the post-set of a place is one for every possible configuration is defined similarly, but using the terms $\Phi_{(p,t_i)}$ in the lifted post-set of p ($p°$).

This way, a PNPL includes some Petri net that is a MG if there is a feature configuration ρ such that for every place p in the PNPL:

- p is not in PN_ρ, therefore Φ_p is false; or
- p is in PN_ρ, and therefore $\Phi_{°p}$ and $\Phi_{p°}$ need to be true.

We can express these conditions as the logical formula in Equation 2.

$$
\Phi_{MG} = \wedge_{p \in P}[\neg\Phi_p \vee (\Phi_p \wedge \Phi_{°p} \wedge \Phi_{p°})]
\tag{2}
$$

If $SAT(\Psi \wedge \Phi_{MG})$ is true, then the PNPL is a weak MG. In such a case, we can use a constraint solver to obtain a feature configuration that satisfies the formula. The Petri net derived using this feature configuration is ensured to be a MG.

Conversely, the feature configurations making the formula Φ_{MG} false yield Petri nets that are not MGs. Hence, a PNPL is a strong MG if $SAT(\Psi \wedge \neg\Phi_{MG})$ is unsatisfiable (i.e., no valid configuration produces a net that is not a MG).

Example. In the PNPL consisting of the feature model in Fig. 2 and the 150% net in Fig. 3, the interesting cases are those for places in and ctrl. In the latter case, any Petri net that contains either both transitions inc_1 and inc_2, or both transitions prod and fix, is not a MG because place ctrl would have either two incoming or two outgoing arcs. This is the case for the Petri nets derived from configurations that select the features Parallel or QualityControl. Similarly, place in will have two incoming arcs for configurations that select the feature Quality-Control, and two outgoing arcs for configurations that select the feature Parallel, resulting in nets that are not MGs. Overall, the example PNPL is not a strong MG product line. However, it is a weak MG product line as, for example, the configuration that only selects features PartA and Prod1 produces a Petri net that is a MG. In practice, if we would like to have no conflicts in the flexible assembly line, we might rule out the problematic variants (i.e., those that are not MGs) by extending the formula Ψ in the feature model. The concrete formula, not reduced, corresponding to the MG property of our example PNPL is the following:

$$
\begin{aligned}
\Phi_{MG} = {} & (\neg PartA \vee (PartA \wedge PartA \wedge PartA)) \\
& \wedge\ (\neg PartB \vee (PartB \wedge PartB \wedge PartB)) \wedge (QualityControl \wedge Parallel) \\
& \wedge\ (\neg Parallel \vee (Parallel \wedge Parallel \wedge Parallel)) \\
& \wedge\ (Parallel \wedge QualityControl) \wedge (\neg Prod1 \vee (Prod1 \wedge Prod1 \\
& \wedge\ (Prod1 \wedge Prod2))) \wedge (\neg Prod2 \vee (Prod2 \wedge Prod2 \wedge (Prod1 \wedge Prod2))) \\
& \wedge\ ((\neg(Prod1 \wedge Prod2)) \vee ((Prod1 \wedge Prod2) \wedge (Prod1 \wedge Prod2)))
\end{aligned}
$$

Then, to assess the MG property on the PNPL, we analyse the satisfiability of the conjunction of this formula Φ_{MG} and the formula of the feature model Ψ.

Interestingly, the lifted analysis of the MG property is very similar to the analysis of the state-machine property. A state-machine (SM) is a subclass of Petri net where each transition t has exactly one input place and one output place, while each place may have multiple input and output transitions. This way, analysing whether a PNPL is a weak/strong SM product line is dual to checking the MG property in a PNPL but replacing transitions by places and vice versa (details in the Appendix).

3.3   Lifted Analysis of the Free-Choice Property

Next, we define the free-choice (FC) property. In a FC net, it is not possible to mix choice and synchronization into one routing construct, i.e., either a choice is preceded by a synchronization, or vice versa. FC Petri nets do not have conflicts since every transition has a unique input place.

Definition 9 (Free-choice, from [7]). *A Petri net $PN = (P, T, A)$ is a free-choice Petri net, written $PN \models FC$, if for every two transitions t_1 and $t_2 \in T$, $t_1 \neq t_2 : {}^\bullet t_1 \cap {}^\bullet t_2 \neq \varnothing \Rightarrow |{}^\bullet t_1| = |{}^\bullet t_2| = 1$.*

In other words, a Petri net is FC if every place is either connected to a unique output transition, or all its output transitions have a unique input place. Formally:

$$\forall p \in P : |p^\bullet| = 1 \vee \forall t \in p^\bullet : |{}^\bullet t| = 1 \tag{3}$$

Following the rationale of the previous analysis, we first lift the definition of property FC to the product line level. Hence, a PNPL is a strong (weak) FC if all (some) its derivable nets are FC.

Definition 10 (Strong and weak FC product line). *A Petri net product line PNL is a strong free-choice iif $\forall PN_\rho \in Prod(PNL): PN_\rho \models FC$. PNL is a weak free-choice iif $\exists PN_\rho \in Prod(PNL): PN_\rho \models FC$.*

According to Eq. 3, every outgoing arc from a place either is unique, or is the only incoming arc to the target transition of the arc. Therefore, a PNPL includes a FC Petri net if there is a feature configuration ρ such that for every place p in the PNPL:

– p is not in PN_ρ, therefore Φ_p is false; or
– p is in PN_ρ, and therefore either $\Phi_{p\circ}$ is true, or for every transition t in the post-set $p^\bullet$:
  • t is not in PN_ρ, therefore Φ_t is false; or
  • t is in PN_ρ, and therefore $\Phi_{\circ t}$ needs to be true.

Equation 4 shows the encoding of these conditions as a logical formula expressing the cases in which a PNPL is a FC product line.

$$\Phi_{FC} = \wedge_{p\in P}[\neg\Phi_p \vee (\Phi_p \wedge (\Phi_{p\circ} \vee (\wedge_{t\in p^\bullet}[\neg\Phi_t \vee (\Phi_t \wedge \Phi_{\circ t})])))] \qquad (4)$$

If $SAT(\Psi \wedge \Phi_{FC})$ is satisfied, then the PNPL is a weak FC product line. On the contrary, configurations leading to nets that are not FC satisfy $\Psi \wedge \neg\Phi_{FC}$. Therefore, a PNPL is a strong FC product line if $SAT(\Psi \wedge \neg\Phi_{FC})$ does not hold.

Example. The sets of conflicting transitions in the PNPL of Fig. 3 (out$_1$ and out$_2$; prod and fix) only have one input place, and therefore, the example is a strong FC product line. In practice, this means that our example has a sound design: in no variant of our flexible assembly line, synchronization (i.e., sequencing of part production or movement through the conveyors in the assembly line) interferes with conflicts (i.e., choice of paths for parts in the assembly line).

3.4 Lifted Analysis of the Extended Free-Choice Property

Extended-free choice (EFC) Petri nets satisfy a weaker condition than FC Petri nets, and every FC Petri net is also EFC. Informally, we say that a Petri net is EFC if the result of a choice between two transitions is never influenced by the rest of the system. The following definition formalizes this intuition.

Definition 11 (Extended free-choice, from [7]). *A Petri net $PN = (P, T, A)$ is an extended free-choice Petri net, written $PN \models EFC$, if for every two transitions t_1 and $t_2 \in T$, $t_1 \neq t_2 : {}^\bullet t_1 \cap {}^\bullet t_2 \neq \varnothing \Rightarrow {}^\bullet t_1 = {}^\bullet t_2$.*

In an EFC Petri net, if a transition has two or more input places, then all these places must have the same set of output transitions. Formally:

$$\forall t \in T : \forall p_1, p_2 \in {}^\bullet t \ \Rightarrow \ p_1{}^\bullet = p_2{}^\bullet \tag{5}$$

Next, we lift the definition of the EFC property to the product line level. A PNPL is a strong (weak) EFC if all (some) derivable nets are EFC.

Definition 12 (Strong and weak EFC product line). *A Petri net product line PNL is a strong extended free-choice iif* $\forall PN_\rho \in Prod(PNL) : \ PN_\rho \models EFC$. *PNL is a weak extended free-choice iif* $\exists PN_\rho \in Prod(PNL) : \ PN_\rho \models EFC$.

According to Eq. 5, we check that each transition t has the following EFC condition:

- t is not in PN_ρ, therefore Φ_t is false; or
- t is in PN_ρ, and therefore Φ_t needs to be true, and moreover, for every two places p_1 and p_2 in the pre-set ${}^\bullet t$ such that $p_1 \neq p_2$:
  1. each transition t' that is not in the post-set of both p_1 and p_2 is not in PN_ρ, and hence $\Phi(p_i, t')$ is false (for $i = 1$ or 2); and
  2. each transition t' that is in the post-set of both p_1 and p_2 is in PN_ρ (and hence $\Phi(p_1, t')$ and $\Phi(p_2, t')$ are true), or disappears from both post-sets (and therefore $\Phi(p_1, t')$ and $\Phi(p_2, t')$ are false).

In the previous condition, the first requirement demands configurations where the transitions t' that only belong to one of the post-sets (i.e., to $p_1{}^\bullet \setminus p_2{}^\bullet$ or to $p_2{}^\bullet \setminus p_1{}^\bullet$) disappear from this post-set. The second requirement demands the common transitions in $p_1{}^\bullet \cap p_2{}^\bullet$ to be maintained. Equation 6 captures these conditions as a logical formula:

$$\begin{aligned}
\Phi_{EFC(t)} = \neg\Phi_t \vee \big(\Phi_t \wedge_{p_1,p_2 \in {}^\bullet t | p_1 \neq p_2} \big[&\wedge_{t' \in p_1{}^\bullet \setminus p_2{}^\bullet} \ \neg\Phi_{(p_1,t')} \\
&\wedge_{t' \in p_2{}^\bullet \setminus p_1{}^\bullet} \ \neg\Phi_{(p_2,t')} \\
&\wedge_{t' \in p_2{}^\bullet \cap p_1{}^\bullet} \ \Phi_{(p_1,t')} \Leftrightarrow \Phi_{(p_2,t')} \big] \big)
\end{aligned} \tag{6}$$

Therefore, we define $\Phi_{EFC} = \wedge_{t \in T} \Phi_{EFC(t)}$. Consequently, if there exists a feature configuration that satisfies $SAT(\Psi \wedge \Phi_{EFC})$, then there is a derivable Petri net that is EFC, and hence the PNPL is a weak EFC product line. A PNPL is strong EFC if $SAT(\Psi \wedge \neg\Phi_{EFC})$ does not hold.

Example. In the PNPL of Fig. 3, there are two transitions that may have two incoming places in some configurations: proc and pack. However, their incoming places only have those transitions as their output, and therefore, the example is a strong EFC. Actually, as we have seen in Sect. 3.3, the example is a strong FC product line, and in consequence, we can conclude that it is a strong EFC product line as well.

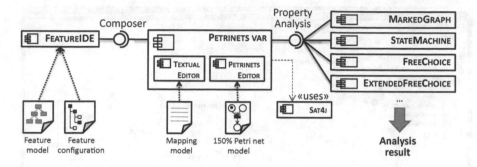

Fig. 6. Architecture of our Petrinets var tool.

4 Tool Support

We have implemented an Eclipse plugin, called Petrinets var, which supports the presented approach. Figure 6 shows its architecture.

Petrinets var provides two editors: one to specify the 150% Petri net, and another to assign PCs to its elements in a so-called mapping model. We use the Eclipse Modeling Framework (EMF) [31] as the underlying modelling technology, and therefore, both the 150% Petri net and the mapping model are EMF-based models that conform to their respective meta-models. The meta-model of the mapping model defines classes to represent the abstract syntax of the boolean formulae making the PCs, together with a cross-reference that points to the Petri net meta-model elements that can be annotated.

We rely on FeatureIDE [19] to specify the feature model and the feature configurations. FeatureIDE provides an extension point Composer that our tool instantiates to automate the derivation of specific Petri nets from the 150% Petri net given a feature configuration. Our tool defines an extension point as well, called Property Analysis, that allows extending the tool with new analysis methods. We currently provide four instances of this extension point to analyse whether some/all Petri nets in a PNPL are state-machines, marked graphs, free-choice or extended free-choice. Since the analysis techniques provided so far rely on the Sat4J solver [5], our plugin provides facilities to transform the conjunction of the analysis formula and the formula of the feature model into conjunctive normal form (CNF), as Sat4J requires. This can simplify the implementation of new analysis techniques by future users.

Figure 7 shows a screenshot of our tool. The Eclipse project explorer (label 1) contains the FeatureIDE project with the definition of the PNPL used as a running example. This project is configured with our composer and declares the 150% Petri net (file *150mm.petrinets*), the feature model (file *model.xml* that is being edited in the window labelled *2*), and the mapping model (file *annotation.vrb* that is being edited in the window labelled *3*). As the figure shows, there are dedicated editors for each kind of file. The textual editor for the PCs has code completion (e.g., offering the available feature names) and validation

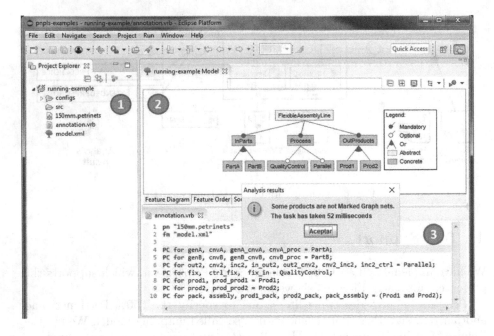

Fig. 7. Screenshot of our Petrinets var tool.

(e.g., it checks that the used features are defined in the feature model). A popup menu on the mapping model allows selecting the lifted analysis to perform. The results of the analyses, including the analysis time, are shown in a dialog box like the one shown in the figure.

Our current implementation uses its own EMF meta-model to represent 150% Petri nets. This meta-model supports a simple notion of net like the one we have used in the paper. However, to improve interoperability with other Petri net tools, we are planning to use the standard Petri Net Markup Language (PNML) [26] instead, for which there is an EMF implementation available.

5   Evaluation

Next, we report on two experiments to assess the efficiency gains of our lifted analyses, compared to generating all derivable nets in a PNPL and analysing each net separately. In the latter case, we perform an explicit analysis of each single net (i.e., without converting the analysis into a constraint satisfaction problem, which may take longer). In our evaluation, we measure the time for analysing the strong satisfaction of the MG, SM, FC and EFC properties (i.e., whether all nets in the PNPL satisfy the property).

We have carried out two experiments based on the running example. In the first experiment, illustrated in Fig. 8, we have analysed the efficiency of our analysis techniques when considering PNPLs with 150% Petri nets of different

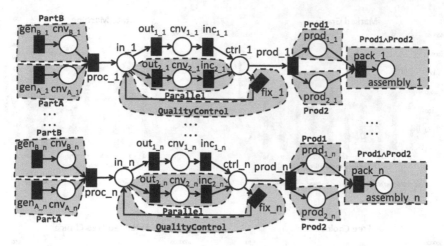

Fig. 8. Experiment 1: PNPL modelling a replicated flexible assembly line.

size but the same number of features. For this purpose, we have created ten PNPLs having the same feature model as in Fig. 2, and whose 150% Petri net contains n replicas of the assembly line in Fig. 3, with n from 1 to 10 (i.e., the first PNPL contains 1 replica, the second one contains two replicas, and so on). All created PNPLs have 36 valid feature configurations. As in the running example, the PNPLs are neither strong MG nor strong SM product lines, but they are strong FC and EFC product lines.

Figure 9 shows the analysis time in milliseconds with logarithmic scale of running 10 times the first execution. We consider just the first execution to discard cache effects. As it can be observed, all lifted analysis techniques were up to three orders of magnitude faster than the time to generate and analyse each net in isolation. In addition, the analysis time did not depend on the size of the 150% Petri net.

The goal of our second experiment is assessing whether an increase in the number of features of the PNPL has an impact in the analysis time. For this purpose, we have created seven PNPLs whose 150% Petri net contains a single assembly line, but they define an increasing number of features to model additional input parts (PartA, PartB, PartC, and so on) and output products (Prod1, Prod2, Prod3, and so on). Figure 10 illustrates the construction of the different feature models. The simplest PNPL contains one input part and one output product, and the most complex one has five input parts and five output products. These PNPLs are constructed by adding replicas of the PartA and Prod1 regions in the PNPL of Fig. 3. The PNPL with one input part and one output product is both a strong MG and a strong SM, but the remaining PNPLs are not. All PNPLs in the experiment are strong FC and EFC product lines.

The graphics in Fig. 11 show the analysis time in milliseconds with logarithmic scale (vertical axis to the left of the graphics), and the number of configurations of each PNPL (vertical axis to the right). Like in the first experiment, we

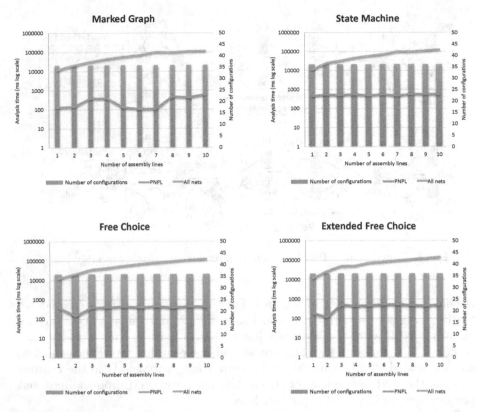

Fig. 9. Analysis time (ms in logarithmic scale) for PNPLs with 150% Petri nets of different size.

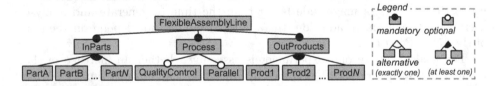

Fig. 10. Experiment 2: PNPL modelling an assembly line with N input parts and output products.

consider just the first execution to discard cache effects. It can be observed that the number of configurations is exponential on the number of features. While the analysis time of all nets in the PNPL one by one is exponential as well, the analysis time of the lifted property is roughly constant. Therefore, the larger the number of features in a PNPL, the bigger the efficiency gains of our lifted analysis compared to an enumerative approach.

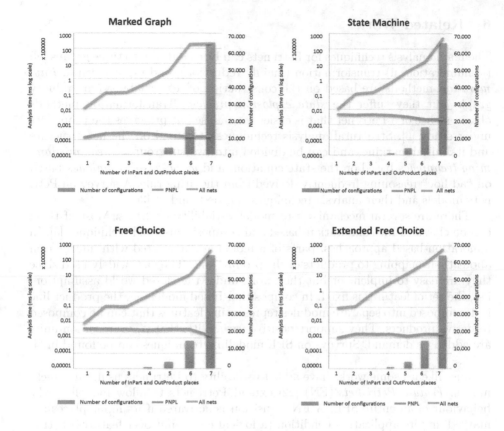

Fig. 11. Analysis time (ms in logarithmic scale) for PNPLs with feature models of different size.

Threats to Validity. While our experiments show gains of at least two orders of magnitude with respect to an enumeration-based approach, the experiments were based on a synthetic net and variations of it generated by replicating either a part of the net or a part of the feature model. Therefore, to further validate these results, we plan to consider models arising in realistic scenarios.

In addition, these results are for checking strong satisfaction of the properties. We plan to extend the experiments to consider weak satisfaction, for which we would require PNPLs with different percentages of product Petri nets satisfying the properties. We expect that our method will provide larger efficiency gains when the percentage of product Petri nets satisfying the property is low, as more nets within the PNPL need to be inspected to find one satisfying the property.

6   Related Work

The main analysis techniques for Petri nets can be classified into three groups [9]: i) enumeration, ii) transformation (mainly reduction), and iii) structural. *Enumeration* methods are based on the construction of a reachability/coverability graph, but they suffer the *state explosion problem*. Transformation methods obtain a slice of a Petri net that is easier to analyse but preserves the properties under study [6]. Structural analysis techniques are based on the net structure and its initial marking, and can be divided into two subgroups: *linear programming techniques* based on the state equation, and *graph-based techniques* based on "ad hoc" reasoning frequently derived from the firing rule. A survey on Petri nets models and their analysis techniques can be found at [30].

There are several mechanisms to model variability for SPLs. Most of them can be classified into annotation-based and composition-based techniques [3]. In annotation-based approaches, parts of a model are annotated with information about their mapping to products of the product line. They are widely used since they are easy to implement but they work under the closed world assumption, i.e., the set of features is fixed. In composition-based modelling, the product line is decomposed into separate modules representing features that can be composed to derive products. They support positive variability, that is, composition units are added on demand. Surveys on SPL modelling techniques can be found in [4, 8,33].

Just like us, some works have added variability to Petri nets using SPL techniques. *Feature Petri nets* (FN) [22] extend Petri nets to allow modelling the behaviour of an entire SPL. A FN transition is activated if its input places are marked and its application condition (a logical constraint over features) is true under the current configuration state. *Dynamic feature Petri nets* (DFPN) [23] extend FN to control feature bindings at runtime, and allow the evaluation of some dynamic properties using model checking. These works lift analysis techniques based on the reachability graph to the product line level, by adding PCs to this graph. They follow an annotative approach to model variability. Similarly, some works have used variability in Petri nets to express variants of higher-level languages – like activity diagrams – and use a variable reachability graph for analysis [14]. This work also uses an annotative approach for SPLs. With respect to these works, our mapping model is more general: [14] only supports variability in edges and [23] only supports variability in arcs and transitions, while our approach permits PCs in arcs, places and transitions. With respect to analysis techniques, the mentioned works focus on the reachability graph, while we lift structural analysis techniques.

In addition to SPL methods, other techniques to handle variability in Petri nets have been proposed. *Conditional Petri nets* [34] associate to each transition a condition defined with the family of $\mathcal{L}$ languages, and transitions are conditioned by the transition sequence previously applied. Likewise, *logical Petri nets* [17] limit transition firing by means of constraints on first-order logic. There are also *reconfigurable nets* [18], which can change the net topology at runtime by means of rewriting rules. Instead, PNPLs are static: the user needs to provide a configuration to derive a Petri net.

Regarding analysis of model-based product lines, Czarnecki and Pietroszek [10] propose an approach to check whether all possible derivable models satisfy the OCL constraints of their meta-model. We may have encoded the different structural properties in OCL and used that technique. However, our solution permits generating specific constraints for the analysed PNPL (instead of relying on one generic OCL constraint), which therefore can be solved using simpler and potentially more efficient standard SAT-solving techniques. Instead, Czarneck's approach requires extending an existing OCL-based checker to consider PCs, while in practice we just use the Sat4J SAT solver.

Concerning SPL analysis of temporal properties, Legay et al. [16] represent the behaviour of variability-intensive systems by means of an extension of transition systems, called Feature Transition System. These authors also propose model checking algorithms to verify all products of a SPL [8]. Unlike this approach, we only focus on static properties, but we plan to explore behavioural properties in future works.

Altogether, to the best of our knowledge, there are no previous works on lifting the analysis of structural properties of PNPLs. This is relevant to enable an efficient analysis of structural properties for all variants within a PNPL. Structural analysis helps in discovering possible design errors in some product Petri net, e.g., related to the existence of conflicts, or interference of synchronization with conflicts. Our work is a first step in this direction, which we have realized in practice through extensible tooling.

7   Conclusions and Future Work

In this paper, we have proposed the notion of Petri net product line, and showed how to analyse structural properties (the marked graph, state-machine, free choice and extended free choice properties) at the product line level. We have validated the approach in practice by presenting an extensible prototype on top of FeatureIDE, and an experiment that shows the benefits of our approach with respect to an enumerative one.

In the future, we plan to support more types of static analysis techniques, exploit compositionality of Petri nets in these analysis techniques, and perform more thorough experiments. We also plan to consider further types of properties (not only strong and weak), in the line of [13]. Our idea is to develop a domain-specific language to express such analyses, which then can be compiled into standard SAT solving procedures. At the tool level, we will use the PNML meta-model to ease the connection of our approach with Petri net tools like CPN Tools [35]. We are also planning to explore the lifting of dynamic analysis techniques, e.g., based on the incidence matrix and on the reachability graph (similar to [23]). For the former, our idea is to include the PCs on the elements of the matrix, and use constraint solving to find place/transition invariants for some/all Petri net products. For the latter, a first idea – if we restrict to Petri

nets with PCs only in transitions – is to calculate the reachability graph of the
150% Petri net, and then annotate the reachability graph with PCs, in the style
of [23]. Finally, we would like to combine product lines with other types of Petri
nets (e.g., with inhibitor or read arcs, or with timed transitions), and consider
variability of the Petri net language itself.

Acknowledgments. Work funded by the Spanish Ministry of Science (RTI2018-
095255-B-I00) and the R&D programme of Madrid (P2018/TCS-4314).

Appendix

This appendix lifts the analysis of the state-machine property. A state-machine
(SM) is a subclass of Petri net where each transition t has exactly one input
and one output place, while each place may have multiple input and output
transitions. SMs allow representing decisions, but not the synchronization of
concurrent activities.

Definition 13 (State-machine, from [21]). *A Petri net $PN = (P,T,A)$ is a
state-machine, written $PN \models SM$, if $\forall t \in T: |{}^\bullet t| = |t^\bullet| = 1$.*

Next, we lift the definition of the SM property to the product line level. A
PNPL is a strong (weak) SM if all (some) derivable nets are SMs.

Definition 14 (Strong and weak SM product line). *A Petri net product
line PNL is a strong state-machine iif $\forall PN_\rho \in Prod(PNL): PN_\rho \models SM$.
PNL is a weak state-machine iif $\exists PN_\rho \in Prod(PNL): PN_\rho \models SM$.*

Similar to the case of MGs, to ensure that all derivable nets are SMs, we
check that the size of the lifted pre-set ${}^\circ t = \{(p_0, \Phi_{(p_0,t)}), ..., (p_n, \Phi_{(p_n,t)})\}$ and
the lifted post-set $t^\circ = \{(p_0, \Phi_{(t,p_0)}), ..., (p_n, \Phi_{(t,p_n)})\}$ of a transition t is one in
every configuration. The size of the lifted pre-set ${}^\circ t$ of a transition t is one if the
following formula is true. The disjunction starts with $false$ to consider the case
when ${}^\circ t$ is empty. The formula Φ_{t° to check that the size of the lifted post-set of
a transition t is one is defined similarly.

$$\Phi_{\circ t} \triangleq false$$
$$\vee (\Phi_{(p_0,t)} \wedge \neg\Phi_{(p_1,t)} \wedge ... \wedge \neg\Phi_{(p_n,t)})$$
$$\vee (\neg\Phi_{(p_0,t)} \wedge \Phi_{(p_1,t)} \wedge ... \wedge \neg\Phi_{(p_n,t)}) \tag{7}$$
$$\vee ... (\neg\Phi_{(p_0,t)} \wedge \neg\Phi_{(p_1,t)} \wedge ... \wedge \Phi_{(p_n,t)})$$

Hence, a PNPL includes some Petri net that is a SM if there is a feature
configuration ρ such that for every transition t in the PNPL:

- t is not in PN_ρ, therefore Φ_t is false; or
- t is in PN_ρ, and therefore $\Phi_{\circ t}$ and Φ_{t° need to be true.

Equation 8 shows the formula that captures the two previous conditions.

$$\Phi_{SM} = \wedge_{t \in T}[\neg\Phi_t \vee (\Phi_t \wedge \Phi_{\circ t} \wedge \Phi_{t\circ})] \tag{8}$$

If there is a feature configuration such that $SAT(\Psi \wedge \Phi_{SM})$ holds, then there is a derivable Petri net that is a SM, and the PNPL is a weak SM. On the contrary, the feature configurations that produce nets which are not SMs are those making the formula $\neg\Phi_{SM}$ true. Hence, the PNPL is a strong SM if $SAT(\Psi \wedge \neg\Phi_{SM})$ does not hold.

References

1. van der Aalst, W.: Structural characterizations of sound workflow nets. Computing Science Reports 9263, Technische Universiteit Eindhoven (1996)
2. van der Aalst, W., Kindler, E., Desel, J.: Beyond asymmetric choice: a note on some extensions. Petri Net Newsl. **55**, 3–13 (1998)
3. Apel, S., Batory, D., Kästner, C., Saake, G.: Feature-Oriented Software Product Lines - Concepts and Implementation. Springer, Heidelberg (2013). https://doi.org/10.1007/978-3-642-37521-7
4. Benduhn, F., Thüm, T., Lochau, M., Leich, T., Saake, G.: A survey on modeling techniques for formal behavioral verification of software product lines. In: VaMoS, pp. 80:80–80:87. ACM (2015). https://doi.org/10.1145/2701319.2701332, http://doi.acm.org/10.1145/2701319.2701332
5. Berre, D.L., Parrain, A.: The Sat4j library, release 2.2. JSAT **7**(2–3), 59–64 (2010)
6. Berthelot, G.: Transformations and decompositions of nets. In: Brauer, W., Reisig, W., Rozenberg, G. (eds.) ACPN 1986. LNCS, vol. 254, pp. 359–376. Springer, Heidelberg (1987). https://doi.org/10.1007/978-3-540-47919-2_13
7. Best, E.: Structure theory of petri nets: the free choice hiatus. In: Brauer, W., Reisig, W., Rozenberg, G. (eds.) ACPN 1986. LNCS, vol. 254, pp. 168–205. Springer, Heidelberg (1987). https://doi.org/10.1007/978-3-540-47919-2_8
8. Classen, A., Cordy, M., Schobbens, P., Heymans, P., Legay, A., Raskin, J.: Featured transition systems: foundations for verifying variability-intensive systems and their application to LTL model checking. IEEE Trans. Softw. Eng. **39**(8), 1069–1089 (2013)
9. Colom, J., Teruel, E., Silva, M.: Performance Models for Discrete Event Systems with Synchronisations: Formalisms and Analysis Techniques. Ed. KRONOS (1998)
10. Czarnecki, K., Pietroszek, K.: Verifying feature-based model templates against well-formedness OCL constraints. In: GPCE, pp. 211–220. ACM (2006)
11. Desel, J., Esparza, J.: Free Choice Petri Nets. Cambridge University Press, Cambridge (1995)
12. Gómez-Martínez, E., de Lara, J., Guerra, E.: Towards extensible structural analysis of Petri net product lines. In: PNSE, vol. 2424, pp. 37–46. CEUR (2019)
13. Guerra, E., de Lara, J., Chechik, M., Salay, R.: Property satisfiability analysis for product lines of modelling languages. IEEE Trans. Softw. Eng. (2020, in press). https://doi.org/10.1109/TSE.2020.2989506
14. Heuer, A., Stricker, V., Budnik, C.J., Konrad, S., Lauenroth, K., Pohl, K.: Defining variability in activity diagrams and Petri nets. Sci. Comput. Program. **78**(12), 2414–2432 (2013)

15. Kang, K., Cohen, S., Hess, J., Novak, W., Peterson, A.: Feature-oriented domain analysis (FODA) feasibility study. Technical report. CMU/SEI-90-TR-021, Software Engineering Institute, Carnegie Mellon University (1990)
16. Legay, A., Perrouin, G., Devroey, X., Cordy, M., Schobbens, P.-Y., Heymans, P.: On featured transition systems. In: Steffen, B., Baier, C., van den Brand, M., Eder, J., Hinchey, M., Margaria, T. (eds.) SOFSEM 2017. LNCS, vol. 10139, pp. 453–463. Springer, Cham (2017). https://doi.org/10.1007/978-3-319-51963-0_35
17. Liu, W., Wang, P., Du, Y., Zhou, M., Yan, C.: Extended logical Petri nets-based modeling and analysis of business processes. IEEE Access 5, 16829–16839 (2017)
18. Llorens, M., Oliver, J.: Structural and dynamic changes in concurrent systems: reconfigurable Petri nets. IEEE Trans. Comput. 53(9), 1147–1158 (2004). https://doi.org/10.1109/TC.2004.66
19. Meinicke, J., Thüm, T., Schröter, R., Benduhn, F., Leich, T., Saake, G.: Mastering Software Variability with FeatureIDE. Springer, Heidelberg (2017). https://doi.org/10.1007/978-3-319-61443-4
20. Meyers, B., Mierlo, S.V., Maes, D., Vangheluwe, H.: Efficient software controller variant development and validation (ECoVaDeVa) overview of a flemish ICON project. In: STAF Co-Located Events, vol. 2405, pp. 49–54. CEUR (2019)
21. Murata, T.: Petri nets: properties, analysis and applications. Proc. IEEE 77(4), 541–580 (1989)
22. Muschevici, R., Clarke, D., Proença, J.: Feature petri nets. In: SPLC Workshops, pp. 99–106. Lancaster University (2010)
23. Muschevici, R., Proença, J., Clarke, D.: Feature nets: behavioural modelling of software product lines. Softw. Syst. Model. 15(4), 1181–1206 (2016). https://doi.org/10.1007/s10270-015-0475-z
24. Nabi, H., Aized, T.: Modeling and analysis of carousel-based mixed-model flexible manufacturing system using colored Petri net. Adv. Mech. Eng. 11(12), 1–14 (2019)
25. Northrop, L., Clements, P.: Software Product Lines: Practices and Patterns. Addison-Wesley Longman Publishing Co., Inc., Boston (2002)
26. Petri Net Markup Language. www.pnml.org
27. Pohl, K., Böckle, G., van der Linden, F.: Software Product Line Engineering. Foundations Principles and Techniques. Springer, Heidelberg (2005). https://doi.org/10.1007/3-540-28901-1
28. Rosa, M.L., van der Aalst, W., Dumas, M., Milani, F.: Business process variability modeling: a survey. ACM Comput. Surv. 50(1), 2:1–2:45 (2017)
29. Seidl, C., Schaefer, I., Aßmann, U.: DeltaEcore - a model-based delta language generation framework. In: Modellierung. LNI, vol. 225, pp. 81–96. GI (2014)
30. Silva, M.: Half a century after Carl Adam Petri's Ph.D. thesis: a perspective on the field. Ann. Rev. Control 37(2), 191–219 (2013). https://doi.org/10.1016/j.arcontrol.2013.09.001
31. Steinberg, D., Budinsky, F., Paternostro, M., Merks, E.: EMF: Eclipse Modeling Framework 2.0, 2nd edn. Addison-Wesley Professional, Boston (2009)
32. Teruel, E., Silva, M.: Structure theory of equal conflict systems. Theoret. Comput. Sci. 153(1&2), 271–300 (1996)
33. Thüm, T., Apel, S., Kästner, C., Schaefer, I., Saake, G.: A classification and survey of analysis strategies for software product lines. ACM Comput. Surv. 47(1), 6:1–6:45 (2014). https://doi.org/10.1145/2580950

34. Tiplea, F., Jucan, T., Masalagiu, C.: Conditional Petri net languages. Elektronische Informationsverarbeitung und Kybernetik **27**(1), 55–66 (1991)
35. Westergaard, M., Kristensen, L.M.: The Access/CPN framework: a tool for interacting with the CPN tools simulator. In: Franceschinis, G., Wolf, K. (eds.) PETRI NETS 2009. LNCS, vol. 5606, pp. 313–322. Springer, Heidelberg (2009). https://doi.org/10.1007/978-3-642-02424-5_19

Stability of Regional Orthomodular Posets Under Synchronisation and Refinement

Federica Adobbati[1], Carlo Ferigato[3], Stefano Gandelli[1], and Adrián Puerto Aubel[2(✉)]

[1] DISCo—Università degli Studi di Milano-Bicocca, Milan, Italy
[2] Inria - Rennes Bretagne-Atlantique, IRISA, Université de Rennes I, Rennes, France
puerto.adrian@gmail.com
[3] JRC—European Commission, Joint Research Centre, Ispra, Italy

Abstract. The *regions* of a *condition/event transition system* can be used to identify the sequential components of the distributed system it represents. With the aim of analysing such a system with respect to its local states, we study the structure obtained from ordering the regions by set inclusion. The resulting algebraic structure is an *orthomodular partial order* (OMP). Given an OMP, one can then define another condition/event transition system, canonical with respect to it. We are interested in characterising cases in which an OMP is *stable*, i.e. it is isomorphic to the OMP obtained as the regional structure of its canonical transition system. We propose, to this aim, a composition operation, and a refinement operation for stable orthomodular partial orders, the results of which are stable.

1 Introduction

This work studies the interrelations between local states, locally observable properties, and events of distributed systems. It extends the work presented in [1]. Our framework is in the relation between Petri net systems, and the labelled transition systems expressing their behaviour, their case graphs.

Labelled transition systems are a widespread class of models, suitable for specifying the desired behaviour of a system. They are commonly used, for instance, to verify that the system complies with a given specification [10]. These models represent systems as their collections of possible states, and the events that lead them from one state to another. Among their features, we focus on their expression of concurrency by means of interleaving semantics. Indeed, interleaving of events is understood as their causal independence, which can in turn be exploited to assign these events to different localities. It is common practice, in particular in the field of Petri net synthesis [2], to let this interleaving guide the discovery of the local states of a system, so as to better represent the way it can be distributed. In this way, the states of the transition system, henceforth referred to as global, can be expressed as particular combinations of the so discovered local states.

© Springer-Verlag GmbH Germany, part of Springer Nature 2021
M. Koutny et al. (Eds.): ToPNoC XV, LNCS 12530, pp. 50–74, 2021.
https://doi.org/10.1007/978-3-662-63079-2_3

We focus on the particular case in which these local states act like Boolean variables, forbidding the execution of events by carrying true or false values. Thus, our framework is narrowed down to elementary and condition/event net synthesis [15], and their close relation to the theory of regions. This relation was first noted by Ehrenfeucht and Rozenberg in [11,12]. In a labelled transition system, a region is a set of global states having a consistent incidence with respect to the events, so as to be the extension of a local state. Such a system is said to be elementary, or condition/event, if it has enough regions to actually guide its behaviour, and it is then said to be separated. In this case, a Petri net system can be constructed with these regions as local states, such that its case graph will be isomorphic to the initially specified transition system. A duality arises: global states can be expressed as (logically) consistent collections of local states, and each local state is identified with a region.

Regions, as subsets of global states of a system, form an algebraic structure [4]. This structure arises naturally when interpreting them as the extensions of locally observable properties of the system. They can be ordered by set inclusion, in a relation that expresses implication. Negation, disjunction, and conjunction can be defined as set operations, only locally though, in the latter two cases. The resulting structure was shown to be an orthomodular partial order [4], OMP for short. Orthomodular partial orders typically represent, in an algebraic fashion, systems about which the acquisition of information depends on the point of view. They were first introduced by Birkhoff and Von Neumann [8] in order to formalise the logic of quantum systems. Foulis and Randall [13,19] extended their approach to provide a general model, referred to as *test space*, which represents the dependencies between the different experiments one can perform on a given system.

When the elements of the OMP are regions, it is convenient to consider observers of the system rather than experiments. Furthermore, each such observer is bound to a spatial location. In this way, they can individually observe only a part of the distributed system, the behaviour observable from their location. The theory of Petri nets justifies that, in this setting, only sequential behaviour is locally observable. In other words, two events will only be concurrent in the system if they do not involve a same location. The OMP then depicts synchronisations as the observations which are common to different localities, and expresses the logical consequences they have on each of the involved local components. This last interpretation justifies our view that OMPs are suitable models for specifying the physical implementation of distributed systems, further allowing for a logical analysis. Such a consideration raises the question of whether any given OMP can be obtained from the interleaving of events of some transition system.

We hence consider a second synthesis problem: Given an OMP determine whether it is the regional structure of some condition/event transition system. This problem was first posed in [4], and a solution was pointed at. This solution involves the construction of a condition/event transition system which is canonical with respect to the OMP. Indeed, when an OMP is *rich*, one can make use

of a representation theorem to find a set of global states such that the elements
of the OMP are subsets of it, in which case we say the OMP admits a concrete
representation.[1] By considering a complete directed graph whose vertices are
these global states, one can label each arc with an event so that all elements
of the OMP can be found as regions of this new transition system, tagged as
saturated. When the structure of regions of the saturated system is isomorphic
to the specified OMP, then the latter is called *stable*. Not every OMP is stable. In
fact, when it is not rich, it does not admit a concrete representation.

The full characterisation of the class of stable OMPs is an open problem,
first posed in [4, p.667] in the setting of category theory, but we here prefer to
restate it in terms of stability. Note that, by construction, any stable OMP must
be isomorphic to a regional OMP, the structured collection of regions of some
elementary transition system. It is further conjectured that every regional OMP
is stable. The present work extends a series of papers [1,4–7] the scope of which
is to study this stability of OMPs, in an attempt to prove the conjecture. In [6],
some classes of stable OMPs were presented. Here, two operations of composition
and refinement are defined, and it is proved that they preserve stability. These
results then allow to show the stability of regional OMPs for a wider range of
systems.

2   Background

This section recalls the different definitions, and established results that will
serve as a ground for the contributions of the present work. We first present the
class of transition systems at stake, as well as the basic facts about orthomodular
posets, while pointing out their relations. The last subsection will lead into our
line of research by presenting the results that launched it in [4].

2.1   Transition Systems and Regions

A labelled transition system is a directed graph whose vertices are called states,
and whose arcs, called transitions, are labelled over an alphabet whose elements
are called events. They are commonly used to specify the behaviour of distributed
systems, and as models on which to verify whether the system satisfies desired
properties [10].

Definition 1 (Transition System). *A transition system is a structure* $A =$
(Q, E, T)*, where Q is a set of states, E is a set of events and $T \subseteq Q \times E \times Q$
is a set of transitions carrying labels in E, such that:*

1. *the underlying graph of the transition system is connected;*
2. $\forall (q_1, e, q_2) \in T \quad q_1 \neq q_2$*;*

---

[1] We remark that our notion of state on OMPs is distinct from the notion of state in
   the space of the configurations of event structures as in [16] since the regions of a
   condition/event transition system are based on the labels associated to the events
   and not to the individual occurrence of events as in occurrence transition systems.

3. $\forall (q, e_1, q_1), (q, e_2, q_2) \in T \quad q_1 = q_2 \Rightarrow e_1 = e_2;$
4. $\forall e \in E \quad \exists (q_1, e, q_2) \in T.$

All the structures we here consider are finite. It will be relevant to consider the following postulates, which guide our interpretation. First, the principle of extensionality [18] states that events are characterised by their observable effect, and can thus only be distinguished from one another by an observation of their effect on the system. Second, the principle of locality states that all observations are attached to a spatial location by the laws of physics, so that only an ubiquitous observer could globally apprehend a distributed system at a given instant. According to these, an event must be characterised by the modification, if any, that it performs on the states of each of the local components of the system. Conversely, when the identity of the events is coherently provided, as for example in a labelled transition system, then these principles allow for inferring the different locations of the system, assuming it is maximally distributed over space. This is achieved thanks to the notion of region.

Regions were introduced in [11,12] and, in several variants, are the fundamental tool for solving the so called *synthesis problem* for Petri nets.

Definition 2 (Region). *A region of a transition system* $A = (Q, E, T)$ *is a subset* r *of* Q *such that:* $\forall e \in E, \forall (q_1, e, q_2), (q_3, e, q_4) \in T$:

1. $(q_1 \in r$ *and* $q_2 \notin r)$ *implies* $(q_3 \in r$ *and* $q_4 \notin r)$ *and*
2. $(q_1 \notin r$ *and* $q_2 \in r)$ *implies* $(q_3 \notin r$ *and* $q_4 \in r)$.

Regions are subsets of states which have a consistent orientation with the labelling of transitions. As such, they can be assigned an incidence with respect to each event, and interpreted as Boolean conditions which hold at the states composing them. In this way, events may be described by the change of truth values they apply to these conditions. In the following, we will say that a region is the *extension* of the corresponding condition, or property. In general, the extension of a property is the collection of states at which it holds, and the property is considered observable when its extension is indeed a region.

Example 1. The right-hand side of Fig. 1 shows a labelled transition system. The set $\{q_1, q_2\}$ is a region, all occurrences of e_1 exit it, and the only occurrence of e_4 enters it. All occurrences of each of the remaining events are independent from it, they do not cross the border of its extension.

The set of regions of a transition system A will be denoted by $\mathcal{R}(A)$, and given a state q of A, R_q will denote the subset of the regions in $\mathcal{R}(A)$ which contain the state q. Given a transition system A and its set of regions $\mathcal{R}(A)$, the *pre-set* and *post-set* operators on the regions and on the *events* of A will be used in the main part of this contribution.

Definition 3 (Pre- and Post-sets). $^\bullet(.)$, *the pre-set operator, and* $(.)^\bullet$, *the post-set operator are defined on the events* E *and on the regions* $\mathcal{R}(A)$ *of the transition system* A *as follows. Let* r *be a region in* $\mathcal{R}(A)$:

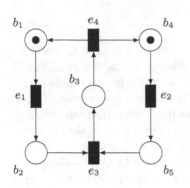
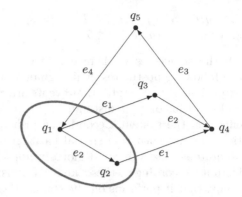

Fig. 1. To the left, a Condition/Event net system. To the right, the transition system corresponding to its case graph. To the right a subset of states is drawn. It corresponds to the extension of b_1: $\{q \in Q \mid b_1 \in q\} = \{q_1, q_2\}$.

1. $^\bullet r = \{e \in E \mid \exists\, (q_1, e, q_2) \in T \text{ such that } q_1 \notin r \text{ and } q_2 \in r\}$;
2. $r^\bullet = \{e \in E \mid \exists\, (q_1, e, q_2) \in T \text{ such that } q_1 \in r \text{ and } q_2 \notin r\}$;
3. $^\bullet e = \{r \in \mathcal{R}(A) \mid e \in r^\bullet\}$;
4. $e^\bullet = \{r \in \mathcal{R}(A) \mid e \in {}^\bullet r\}$.

We are concerned with a specific class of transition systems, the *condition/event transition systems* (CETS for short). CETS can be defined as those labelled transition systems which are the case graph of some condition/event net system. However, the characterisation of these, in terms of their regions [11,12], allows us to provide an alternative definition, in terms of the so called separation axioms. It follows from the results of [11,12] that the two definitions are equivalent.

Definition 4 (CETS – Condition/Event Transition System). *A* CETS *is a transition system $A = (Q, E, T)$, in which the following conditions hold:*

1. $\forall\, q_1, q_2 \in Q \quad R_{q_1} = R_{q_2} \Rightarrow q_1 = q_2$;
2. $\forall\, q_1 \in Q \,\forall e \in E \quad {}^\bullet e \subseteq R_{q_1} \Rightarrow \exists\, q_2 \in Q \text{ such that } (q_1, e, q_2) \in T$;
3. $\forall\, q_1 \in Q \,\forall e \in E \quad e^\bullet \subseteq R_{q_1} \Rightarrow \exists\, q_2 \in Q \text{ such that } (q_2, e, q_1) \in T$.

This definition requires of a transition system, that its set of regions is sufficient to fully determine the incidence of each event with respect to each state. In this way, events are not only described, but fully characterised by the truth-value modifications they apply to the observable properties of the system. Under these conditions, a condition/event net system can be constructed, which has the set of regions for conditions, and the transition labels as events, and such that its case graph is isomorphic to the original transition system. This tight relation between condition/event net systems, and CETS was thoroughly formalised in [15], in the equivalent framework of elementary systems.[2]

---

[2] For this reference, please note that in the elementary framework, the definition of region coincides with the one we give here. Elementary transition systems and CETS only differ in state reachability considerations.

The following example is meant to provide some intuition to the readers familiar with Petri net theory. The formal definitions regarding Petri nets are out of the scope of this work, and this example is not fundamental for understanding the rest of it.

Example 2. Given the condition/event net system in Fig. 1, we can represent its behaviour through its case graph, in the same figure. The global state $\{b_1, b_4\} = q_1$ is reached only through the occurrence of e_4, when the system is in the state $\{b_3\} = q_5$. From the state q_1, the net system can reach the state $\{b_2, b_4\} = q_3$, if e_1 occurs, or $\{b_1, b_5\} = q_2$, if e_2 occurs. If both e_1 and e_2 occur, the system is in the state $\{b_2, b_5\} = q_4$. Finally, the state q_5 is only reached by firing e_3, starting from the state q_4.

We now show that the opposite construction is also possible, i.e. starting from the CETS on the right of Fig. 1, we can derive the net system on the left through a synthesis procedure. The first step is computing the regions of the CETS. These will represent the conditions of the net, while the labels on the arcs are translated into events. The following sets of states are regions in the CETS: $\{q_1, q_2\} = b_1$, $\{q_3, q_4\} = b_2$, $\{q_2, q_4\} = b_5$, $\{q_1, q_3\} = b_4$ and $\{q_5\} = b_3$. The set of these regions satisfies the list of axioms in Definition 4, therefore they are sufficient for the synthesis of the CETS [2]. For every region, we have to consider which are the events in its pre-set and which are in its post-set. Let us consider the region $\{q_1, q_2\} = b_1$, all the transitions labelled e_1 go out from the region, while the transition e_4 comes into it. In the synthesised net, this is represented by putting b_1 in the post-set of e_4 and in the pre-set of e_1. Analogously, we can determine the relations between all the other conditions and events of the net considering the pre-set and post-set of the other regions listed above.

2.2 Orthomodular Partially Ordered Sets

Orthomodular partially ordered sets, indicated as OMP in what follows, are well known algebraic structures in the literature on *Quantum Logics*. The following definition can be found as Definition 1.1.1 in [17]. Note that $\wedge$ and $\vee$ are the usual meet and join operations on a partial order, denoting respectively the *greatest lower bound*, and the *least upper bound*.

Definition 5 (OMP - Orthomodular Partially Ordered Set). *An ortho-modular partially ordered set $\langle L, \leq, (.)', 0, 1 \rangle$ is a set L endowed with a partial order $\leq$ and a unary operation $(\,.\,)'$ (called orthocomplement), such that the following conditions are satisfied:*

1. *L has a least and a greatest element (respectively 0 and 1) and $0 \neq 1$;*
2. *$\forall x, y \in L : x \leq y \Rightarrow y' \leq x'$*   *$(\cdot)'$ is antitone;*
3. *$\forall x \in L : (x')' = x$*   *$(\cdot)'$ is involutive;*
4. *$\forall x, y \in L : x \leq y' \Rightarrow x \vee y \in L$*   *finite orthogonal joins;*
5. *$\forall x, y \in L : x \leq y \Rightarrow y = x \vee (x' \wedge y)$*   *orthomodular law.*

Two elements x and y in L are said to be orthogonal, *noted by $x \perp y$, whenever $x \leq y'$.*

Axioms 2 and 3 constitute the definition of orthocomplement for $(\cdot)'$. Whenever $(\cdot)'$ is an orthocomplement then $\perp$ is a symmetric relation. Note that $\wedge$ and $\vee$ operations are not required to be globally defined. However, Axiom 4 imposes the existence of a least upper bound for each finite collection of pairwise orthogonal elements. We will often denote $\langle L, \leq, (.)', 0, 1 \rangle$ simply by L, and assume that an OMP L_i has underlying structure $\langle L_i, \leq_i, (.)', 0_i, 1_i \rangle$.

Example 3. The regions $\mathcal{R}(A)$ of a CETS A are shown in [3] to satisfy:

1. $\emptyset, Q \in \mathcal{R}(A)$;
2. $r \in \mathcal{R}(A) \Rightarrow Q \setminus r \in \mathcal{R}(A)$;
3. $r_1, r_2 \in \mathcal{R}(A) \Rightarrow (r_1 \cap r_2 \in \mathcal{R}(A) \Leftrightarrow r_1 \cup r_2 \in \mathcal{R}(A))$.

These properties imply that $\mathcal{R}(A)$ can be interpreted as an OMP whenever A is a CETS. $\emptyset$ and Q serve as 0 and 1 respectively, the order is given by set containment and orthocomplement by set complement; moreover, two regions are *orthogonal* whenever their intersection is empty, and so their union is again a region. Note that, since orthogonal regions have no state in their intersection, they are interpreted as mutually exclusive properties, or conditions.

Throughout this work, the OMP obtained from the regions of a CETS $A = (Q, E, T)$ will be denoted $H(A) = (\mathcal{R}(A), \subseteq, \emptyset, Q, (\cdot)')$, $(\cdot)'$ being the *set complement* operation. An OMP L is called *regional*, when there exists a CETS A such that $H(A)$ is isomorphic to L.

We introduce the notion of compatibility, which is central in the theory of OMPs, and in our case of study. This notion will help define Boolean algebras as a particular class of OMPs, which are most relevant for the analysis of sequential behaviour or, equivalently, locality of observations.

Definition 6 (Compatibility, Boolean Algebra). *Let L be an OMP. We say two elements $x, y \in L$ are* compatible, *and write $x \mathrel{\$} y$ whenever there exist three pairwise orthogonal elements $x_0, z, y_0 \in L : x_0 \perp z \perp y_0 \perp x_0$ such that $x = x_0 \vee z$ and $y = y_0 \vee z$. Equivalently, we may say that $x \mathrel{\$} y$ whenever they share a* common *orthogonal basis $\{x_0, z, y_0\}$. We say two elements are* incompatible, *denoted $x \not\mathrel{\$} y$, when they are not compatible. L is a* Boolean algebra *whenever all of its elements are pairwise compatible, $\forall x, y \in L : x \mathrel{\$} y$.*

Ordered elements and orthogonal elements are always compatible.

Boolean algebras are usually defined as *bounded distributive orthocomplemented lattices*, and it is rather routinary to verify that these definitions are equivalent (see [14, 17]). We often name an OMP B, or B_i when it is required to be a Boolean algebra, knowing that any OMP noted L, L_i might as well be one.

Example 4. The trivial OMP $\mathbf{2} = \{0, 1\}$, with $0 \leq 1$ and $0' = 1$ is a Boolean algebra. Indeed, it has only two elements, and $0 \perp 1$. Using the notation of Definition 6, we can verify that $0 \mathrel{\$} 1$, by putting $x_0 = 0, z = 0, y_0 = 1$.

The next simpler example of Boolean algebra is depicted on top of Fig. 4, as $I = \{0, x, x', 1\}$, where x' denotes the orthocomplement of x.

Given a CETS A, then its OMP of regions $\mathcal{R}(A)$ is a Boolean algebra if, and only if, A has no interleaving of events whatsoever, and thus no two events are concurrent. In this case A can not be distributed. As a matter of fact, two properties must be observable from a same location whenever their extensions are compatible regions [4].

A subset M of an OMP L with the order relation and the orthocomplement operation inherited from L is a sub-OMP of L in case M is an OMP ([17], definition 1.2.2). When a sub-OMP M of L is a Boolean algebra, it is simply called a *Boolean subalgebra*. M is furthermore a *maximal Boolean subalgebra*, when it is a sub-OMP of no other Boolean subalgebra of L than itself. Each element of an OMP belongs at least to one of its maximal Boolean subalgebras. Maximal Boolean subalgebras of a regional OMP represent the sequential components of the underlying system. They can be identified with the different locations on which it may be maximally distributed. Note that a region might belong to different maximal Boolean subalgebras, in which case it corresponds to a property which is observable and modifiable from the different corresponding locations.

In order to formalise the composition of OMPs, as well as to define the notion of *state*, we will require the following definition.

Definition 7 (OMP-morphism). *([17], finite case of definition 1.2.7) Let L_1 and L_2 be OMP. A mapping $f : L_1 \to L_2$ is a morphism of OMPs if the following conditions are satisfied:*

1. *$f(0_1) = 0_2$;*
2. *$\forall x \in L_1 \ f(x') = f(x)'$;*
3. *for any finite sequence $\{x_i \mid i \in I\}$ of mutually orthogonal elements in L_1,*

$$f(\bigvee_{i \in I} x_i) = \bigvee_{i \in I} f(x_i)$$

A morphism $f : L_1 \to L_2$ is an isomorphism *if f is injective, maps L_1 onto L_2 and f^{-1} is a morphism. An isomorphism $f : L_1 \to M$, when M is a sub-OMP of L_2, is called* embedding.

Morphisms of OMPs preserve *compatibility*, order and orthogonality [4].

Two-valued states of an OMP are a well-known concept in the literature on *Quantum Logics* ([17], Definition 2.1.1). We here provide a simpler definition, and refer to *two-valued states* simply as *states*, since the states of a CETS correspond to (two-valued) states of its regional OMP.

Definition 8 (State of an OMP). *Let L be an OMP, and* **2** *be the trivial Boolean algebra* $\mathbf{2} = \{0, 1\}$, *with $0 \leq 1$ and $0' = 1$. A state of L is an OMP-morphism $s : L \to \mathbf{2}$.*

A state of an OMP L is thus a consistent assignment of truth values to the elements of L. As such, it is common to identify it with its *support*: $\{x \in L \mid s(x) = 1\}$. Indeed, each state is the characteristic function of its support.[3] The

---

[3] Note that in [4] the support of a state was called filter, and indeed, its restriction to each maximal Boolean subalgebra yields an order-theoretic prime, principal and maximal filter.

collection of all states of an OMP L will be denoted by $S(L)$. It will also be useful to consider the collection of states which contain a given element. Given an element $x \in L$, let us define the notation $S_x = \{s \in S(L) \mid x \in s\}$. It is worth noting that if we fix a state of a CETS, then the morphism assigning 1 to the regions that contain it is indeed a state of its regional OMP.

Example 5. Let $A = (Q, E, T)$ be a CETS, and $H(A) = (\mathcal{R}(A), \subseteq, \emptyset, Q, (\cdot)')$, be its regional OMP. Then for any $q \in Q$, the OMP-morphism

$$s_q : H(A) \to \mathbf{2} \quad \text{such that} \quad s_q(r) = \begin{cases} 1 & q \in r \\ 0 & q \notin r \end{cases}$$

is a state of $H(A)$.

In the cases at stake in this work, finiteness of both CETSs and OMPs provides particular importance to the notion of *atom*, that will provide a useful characterisation of states.

Definition 9 (Atom of an OMP). *Let L be an* OMP. *An element $a \in L$ is called an atom when it* covers 0, *formally:*
$a \neq 0$ *with* $\forall x \in L : (x \leq a) \Rightarrow (x = a \vee x = 0)$.

The collection of atoms of L is noted $\mathcal{A}(L)$. Straightforward verification shows that two atoms are compatible if, and only if, they are orthogonal. Indeed, when referring to a CETS A, the atoms of $H(A)$ faithfully represent the local states of the system in that they are mutually exclusive if and only if they belong to a same maximal Boolean subalgebra—a same sequential component of the system. This fact provides atoms with particular importance in Petri net synthesis.[4] Their importance relies on the fact that finite OMPs are faithfully represented by the orthogonality relation among their atoms. Each element of L can be retrieved as the join of a collection of pairwise orthogonal atoms. The reader is referred to [7] for the details, and to [13,19] for a full characterisation of this representation of OMPs. Note that, when restricted to atoms, the maximal Boolean subalgebras are simply the maximal cliques of $\perp$.

Remark 1. As is common in an order-theoretic framework, we graphically represent OMPs by their *Hasse diagrams*. In these, order is represented upwards, and only the covering relation $(x \prec y \Leftrightarrow \forall z \in L \setminus \{y\} : (z \leq y \Rightarrow z \leq x))$ is explicitly depicted. This type of representation, however, grows unreadable very quickly, justifying the use of a simpler graphical representation. If L is finite, we may depict only its atoms, and draw solid lines to represent its maximal Boolean subalgebras. In this way, two atoms linked by a solid line are orthogonal, and their join is ensured to exist. This representation is called *block diagram* or *Greechie diagram* as in [20, p.107] or [17, def. 2.4.6]. In Fig. 2 the *Hasse diagram* and the *Greechie diagram* of the same OMP are represented. In either case, two OMPs with identical representation will certainly be isomorphic.

---

[4] In [3], as minimal regions, the atoms of a regional OMP were shown to suffice for solving the synthesis problem.

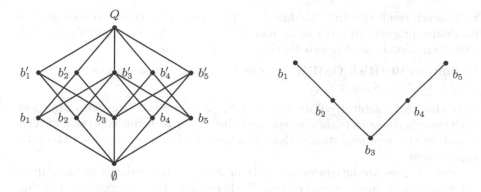

Fig. 2. Two representations of the OMP L obtained by ordering by set containment the regions of the transition system in Fig. 1. To the left, all regions are represented in a *Hasse diagram*, order is represented upwards. To the right, only atoms of the two maximal Boolean subalgebras in L are depicted (see Remark 1).

Since, as said above, we are interested in the specification of a finite OMP L by its collection of *atoms*, $\mathcal{A}(L)$, we will use here a characterisation of states introduced in [4] (proposition 29, p. 649).[5]

Proposition 1. *Let $\mathcal{A}(L)$ be the set of atoms of an OMP L, E the set of the maximal cliques of $\perp$ in L restricted to $\mathcal{A}(L)$, and C the set of the maximal cliques of $ in L restricted to $\mathcal{A}(L)$. Consider $S = \{s \in C \mid \forall e \in E \mid s \cap e \mid = 1\}$, then $s \in S$ iff its up-closure $\uparrow(s) = \{x \in L \mid \exists a \in s : a \leq x\}$ is the support of a state in L.*

This last result states that each state of an OMP selects exactly one atom in each maximal Boolean subalgebra, and that conversely, such a collection of atoms always corresponds to a single state. Since this correspondence is one to one, we will often refer to these particular collections of atoms as the states they characterise, when clear from the context. When interpreting atoms as local states of the system, this correspondence is the simple statement that global states are formed by selecting exactly one local state per location. We remark that, as in [17, prop. 2.4.9], there exist finite OMPs whose set of states is empty. In these cases, the set S in Proposition 1 is empty. The following section will deal with OMPs having "enough" states, so as to be representable in terms of these.

2.3   Saturated Transition System and the Stability Problem

This section is concerned with the representation of an arbitrary OMP as the collection of regions of some CETS. This is not always possible, and in fact, a

---

[5] In [4] a different but equivalent terminology is used: *transitive partial Boolean algebra* for OMP and *filter* for *state*.

well known result due to S. Gudder (see [17]) states that OMPs need to verify a particular property in order to be representable as a collection of subsets, such that their containment represents the order of the OMP.

Definition 10 (Rich OMP). *An* OMP *L is called* rich *whenever*
$\forall x, y \in L : S_x \subseteq S_y \Rightarrow x \leq y.$

Note that in an arbitrary OMP $x \leq y \Rightarrow S_x \subseteq S_y$, and so rich OMPs are those with enough states to faithfully represent their order as inclusion of sets of their states. In this representation orthocomplement is simply given by set-theoretic complement.

However, we are interested not only in interpreting elements of an OMP as subsets of states, but as actual regions. We here report the canonical construction of a transition system from an OMP, first introduced in [4].

Definition 11 (Saturated Transition System). *Let L be an* OMP. *Consider $S(L)$ to be the support of its states. Define the* symmetric differences *between supports of states as $[s, s'] = (s \setminus s', s' \setminus s)$ for $s, s' \in S(L)$, and let $E(L) = \{[s, s'] \mid s, s' \in S(L)\}$, and $T(L) = \{(s, [s, s'], s') \mid s, s' \in S(L)\}$. Then the* saturated transition system *of L is $J(L) = (S(L), E(L), T(L))$.*

Note that each event $e \in E(L)$ will be of the form $e = [s, s'] = (s \setminus s', s' \setminus s)$, as such, it is characterised by its sets of pre-conditions, and post-conditions seen as elements of L. The underlying graph of $J(L)$ is complete, presenting a transition $(s, [s, s'], s')$ between each ordered pair of distinct states (s, s'). In [7] it is shown that the construction of $J(L)$ can be done by considering exclusively the *atoms* of L. In particular, states are determined by their atoms, and so are events, as symmetric differences of the atoms of states. This is displayed in the following example.

Example 6. From the OMP L in Fig. 3 (on the left), we want to construct the transition system $J(L)$. Note that the figure depicts only the atoms of L. In [7], it was shown that we produce isomorphic transition systems whether we consider either the full OMP or its atoms only. So, for the sake of simplicity, we develop this example by considering the latter case. First, we need to compute all the states of L: $S(L) = \{s_1 = \{b_1, b_4, b_6\},\ s_2 = \{b_1, b_5, b_6\},\ s_3 = \{b_1, b_4, b_7\},\ s_4 = \{b_1, b_5, b_7\}, s_5 = \{b_2, b_4\},\ s_6 = \{b_2, b_5\},\ s_7 = \{b_3, b_6\},\ s_8 = \{b_3, b_7\}\}$. Every element in $S(L)$ is a state of $J(L)$. Every pair of states is connected with a transition labelled by the symmetric difference between the states. For example, the transition starting from s_1 and arriving in s_2 is labelled by $\langle \{b_4\}, \{b_5\} \rangle$ (e_1 in the picture). The other labels shown in Fig. 3 are: $e_2 = \langle \{b_6\}, \{b_7\} \rangle$, $e_3 = \langle \{b_1, b_5\}, \{b_3\} \rangle$ and $e_4 = \langle \{b_1, b_7\}, \{b_2\} \rangle$. In order to have a clearer picture, not all the arcs are represented, but $J(L)$ actually has an arc connecting every pair of nodes in both directions. Considering all the arcs starting from s_2, the following are missing in the figure: the arc to s_1, that is labelled $\langle \{b_5\}, \{b_4\} \rangle$; the arc to s_3, labelled $\langle \{b_5, b_6\}, \{b_4, b_7\} \rangle$; the arc to s_5, labelled $\langle \{b_1, b_5, b_6\}, \{b_2, b_4\} \rangle$; the arc to s_6, labelled $\langle \{b_1, b_6\}, \{b_2\} \rangle$; the arc to s_8 labelled $\langle \{b_1, b_5, b_6\}, \{b_3, b_7\} \rangle$. Analogously, we can compute all the events connecting two nodes in the saturated transition system.

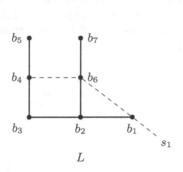

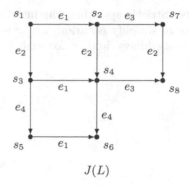

Fig. 3. A OMP L and the associated transition system $J(L)$. In order to have a clearer picture, the transition system represented above is not saturated.

It was shown in [4], that if L is rich, then for each $x \in L$, S_x is a region of $J(L)$, and L (or rather $\{S_x \mid x \in L\}$) is even a sub-OMP of $H(J(L))$. Also, in this case, the regions of $\{S_x \mid x \in L\}$ suffice to satisfy the axioms of Definition 4, so that $J(L)$ is a CETS.

In general, however, $H(J(L))$ may have more elements than L. We hence say that an OMP L is *stable* when it is isomorphic to $H(J(L))$. Note that, by definition, all stable OMPs are indeed regional. It is known that richness is a necessary condition for stability. Some other necessary conditions have been identified in [4–6], and all of these are, in particular, verified by regional OMPs. It is in fact conjectured that all regional OMPs are stable. The following section tackles this conjecture, with an operational approach.

3 Composition of OMPs and Their Stability

This section will start introducing a rather general composition operation for OMPs. The result of this composition will not always be an OMP but we show that it is the case in particular instances. For this composition operation, we show that the results are stable.

3.1 Composition of OMPs

We here present a composition operation for OMPs, in the fashion of those presented in [9] for a less general case.

Definition 12 (V-formation). *A V-formation of OMPs is a tuple* $(I, L_1, L_2, \phi_1, \phi_2)$, *such that* I, L_1, *and* L_2 *are OMPs, and* $\phi_i : I \to L_i (i = 1, 2)$ *are morphisms of OMPs.*

A V-formation serves as specification for composing L_1 and L_2 on the common interface I. In categorical terms, it is simply a diagram in the category of

OMPs. Strictly speaking, the interface would only be $\phi_1^{-1}(L_1) \cap \phi_2^{-1}(L_2)$, so in order to simplify notation, we here consider V-formations in which $\phi_i (i = 1, 2)$ are embeddings, and so for each $i = 1, 2$, $\phi_i(I)$ is a sub-OMP of L_i isomorphic to I.

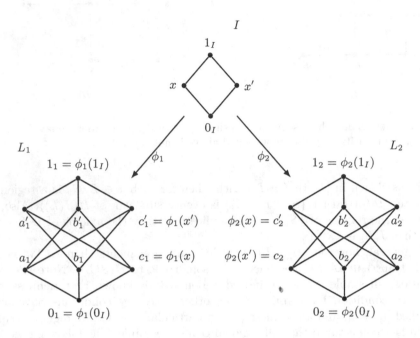

Fig. 4. A V-formation

Given a V-formation of OMPs embeddings, we propose a straightforward composition operation. The nature of the proposed composition interprets the interface I as a sub-OMP of each component, so as to identify the two copies element-wise. In this sense, it requires the morphisms of the V-formation to be actual embeddings.

Definition 13 (Equivalence induced by a V-formation). *Let* $V = (I, L_1, L_2, \phi_1, \phi_2)$ *be a V-formation of* OMP*s. Consider* $\tilde{L}$, *the disjoint union of* L_1 *and* L_2 *such that* $\phi_i : I \to \tilde{L} (i = 1, 2)$, *with* $\phi_1(I) \cap \phi_2(I) = \emptyset$. *The equivalence relation induced by* V *is the binary relation* $\sim_V = \{(x, x) \in \tilde{L}^2 \mid x \in \tilde{L}\} \cup \{(\phi_1(x), \phi_2(x)) \in \tilde{L}^2 \mid x \in I\} \cup \{(\phi_2(x), \phi_1(x)) \in \tilde{L}^2 \mid x \in I\}$.

It is straightforward to verify that $\sim_V$ is an equivalence relation. It is reflexive, and symmetric by construction. If $x, y, z \in \tilde{L}$ are such that $x \sim_V y$ and $y \sim_V z$ then it must be either $z = x$ or $z = y$, so $\sim_V$ is transitive. We will denote the equivalence class of an element $x \in \tilde{L}$ by $[x]$. This equivalence relation satisfies these additional properties:

Proposition 2. *Let* $V = (I, L_1, L_2, \phi_1, \phi_2)$ *be a V-formation of* OMPs, *with* $\tilde{L}$, *and* $\sim_V$ *as in Definition 13. Then:*

1. $\forall i \in \{1, 2\} : \forall x \in L_i : [x] \cap L_i = \{x\}$;
2. $\forall x, y \in \tilde{L} : x \sim_V y \Leftrightarrow x' \sim_V y'$;
3. $[0_1] = [0_2] = \{0_1, 0_2\}$ *and* $[1_1] = [1_2] = \{1_1, 1_2\}$;
4. $\forall x, y \in \tilde{L} : (x \leq y) \Rightarrow \neg(\exists \tilde{x} \in [x], \tilde{y} \in [y] : \tilde{y} < \tilde{x})$.

Proof. 1. By construction.

2. Let $x, y \in \tilde{L}$ be two distinct elements such that $x \sim y$. From the definition of $\sim_V$, it follows that $\exists z \in I : \phi_1(z) = x$ and $\phi_2(z) = y$ (up to swapping of x and y). Since ϕ_1 and ϕ_2 are OMP-morphisms, it follows that $\phi_1(z') = x'$ and $\phi_2(z') = y'$, hence $x' \sim y'$.

3. It is a direct consequence of Axioms 1 and 2 in Definition 7.

4. Let $x \leq y$ then $\exists i \in \{1, 2\} : x, y \in L_i$, and $x \leq_i y$. Analogously, $\exists j \in \{1, 2\} : \tilde{x}, \tilde{y} \in L_j$, and $\tilde{y} <_j \tilde{x}$. Clearly it must be that $i \neq j$. Now, $x \sim_V \tilde{x}$, and $y \sim_V \tilde{y} \Rightarrow \exists a, b \in I : \phi_i(a) = x, \phi_j(a) = \tilde{x}, \phi_i(a) = y$, and $\phi_j(b) = \tilde{y}$. Since ϕ_i is an OMP-embedding, it must reflect the order, yielding $a \leq b$, but ϕ_j preserves the order, and so $\tilde{x} \leq \tilde{y}$.

$\square$

These results allow for endowing the quotient $\tilde{L}/\sim_V$ with a structure.

Definition 14 (*I*-pasting of OMPs). *Consider the setting of Definition 13, and define:*

1. $L = \tilde{L}/\sim_V$,
2. $0 = [0_1] = [0_2]$ *and* $1 = [1_1] = [1_2]$,
3. $\forall [x] \in L : [x]' = [x']$,
4. $[x] \prec [y] \Leftrightarrow \exists x \in [x] : \exists y \in [y] : x \leq y$ *in* $\tilde{L}$, *and*
5. $\leq \subseteq L \times L$ *as the transitive closure of* $\prec$.

Then the I-pasting of L_1 *and* L_2 *induced by V is the structure* $L_1|I|L_2 = \langle L, \leq, (\cdot)', 0, 1 \rangle$.

It follows immediately from Proposition 2 (4.) that $\leq$ is a partial order relation. Proposition 2 (2.), states that $\sim_V$ is congruence for complementation, and so $(\cdot)'$ is well-defined on L. It is furthermore trivial to verify that 0 and 1 are respectively the minimal and maximal elements in L. Moreover, whenever $x \perp y$, then $[x] \perp [y]$.

So defined, the composition of two OMPs over an interface is simply obtained by identifying the elements whose pre-images through ϕ_1 and ϕ_2 coincide.

In general, such an *I*-pasting will be a *partial order* endowed with an operation of *orthocomplement* [14]. This is, however, not sufficient to guarantee that it is in fact an OMP.

The following example shows that it can fail to be.

Example 7. With reference to Fig. 4. Let $I = \{0, 1, x, x'\}$ be an OMP with $0 \leq x, x' \leq 1$. For $i = 1, 2$, let L_i be Boolean algebras with three atoms each $\{a_i, b_i, c_i\}$. Since ϕ_i are OMP-morphisms, $\phi_i(0) = 0_i$, and $\phi_i(1) = 1_i$. Now let $\phi_1(x) = c_1$, so that $\phi_1(x') = c_1' = a_1 \vee b_1$; and $\phi_2(x') = c_2$ so that $\phi_2(x) = c_2' = a_2 \vee b_2$. In this case, $\sim$ is the reflexive and symmetric closure of $\{(0_1, 0_2), (1_1, 1_2), (c_1, c_2'), (c_1', c_2)\}$, let $[x]$ denote its equivalence classes.

In $L_1|I|L_2$, as shown in Fig. 5, we have that $[a_1] \leq [c_1'] = [c_2] \leq [a_2']$. Hence, $[a_1]$ and $[a_2]$ are orthogonal, but they have no least upper bound.

Indeed, $[a_1], [a_2] \leq [b_1'], [b_2']$. This contradicts Axiom 4 of Definition 5.

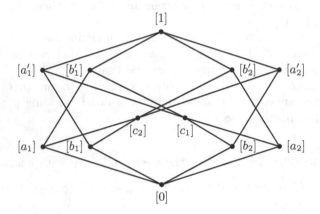

Fig. 5. Hasse diagram of $L_1|I|L_2$ as in Example 7. This composition does not form an OMP.

In what follows, we will study cases in which this composition is actually an OMP.

3.2   Extending a System with a Sequential Component

It is a known result [17] that whenever L_1 and L_2 are OMP's, and $I = \{0, 1\}$ is the trivial Boolean algebra, then $L_1|I|L_2$, as defined in the last subsection, is an OMP. It was further shown in [6] that in this case, if L_1 and L_2 are stable, then so must be $L_1|I|L_2$.

In this work, we consider the case, in which the interface is a non-trivial Boolean algebra $I = \{0, 1, y, y'\}$. In this case $\mathcal{A}(I) = \{y, y'\}$. With such an interface, we impose that the corresponding saturated transition systems synchronise according to the specified embeddings. One of the components will be a finite Boolean algebra, and will be denoted by B. Boolean algebras considered as OMPs were shown to be stable in [6].

The proof of stability will require the notion of *free atom*. An atom is *free* in an OMP if it belongs to just one maximal Boolean subalgebra. Such an atom represents an observation accessible from one single locality, when seen as the region of a CETS, it corresponds to a local state belonging to only one sequential component.

Example 8. With reference to Fig. 2, b_1, b_2, b_4, b_5 and b_3 are all atoms, however b_1, b_2, b_4, b_5 are free, but b_3 is not. Indeed, b_3 belongs to two maximal Boolean algebras.

The considered embeddings will then identify a free atom of an OMP with an atom of B.

Theorem 1. *Let L be an* OMP, *and $x \in \mathcal{A}(L)$ be an atom. Let B be a finite Boolean algebra, and $a \in \mathcal{A}(B)$. Let $I = \{0, 1, y, y'\}$, and define $\phi_L : I \to L$, and $\phi_B : I \to B$ such that $\phi_L(y) = x$, and $\phi_B(y) = a$. Then $L|I|B$ induced by the V-formation $(I, L, B, \phi_L, \phi_B)$ is an* OMP.

Proof. After Proposition 2, it suffices to show that orthogonal joins are well defined, and that the orthomodular law holds. First note that in this case, the only identifications are $[0] = \{0_L, 0_B\}, [1] = \{1_L, 1_B\}, [x] = \{x, a\}$ and $[x'] = \{x', a'\}$, all other equivalence classes being singletons. Since both x and a are atoms, we have that $\leq = \prec$, in the setting of Definition 14. Furthermore, for each pair of orthogonal elements $[c] \perp [d]$, there must be $c \in L' \cap [c]$ and $d \in L' \cap [d]$, where $L' \in \{L, B\}$ such that $c \perp d$ in L'. If this holds for both $L' = L$ and $L' = B$, then the only possibility is $[c] = [x]$ and $[d] = [x']$, for which the join must be $[1]$, and is well defined. Now, from the Definition 14 (4.) of $\prec$, it follows that for every pair of ordered elements $[c] \leq [f]$, there must be one $L' \in \{L, B\}$ such that $c \in L' \cap [c]$ and $f \in L' \cap [f]$, with $c \leq f$. Now, this ensures, on one hand, that orthogonal joins (and meets) are well defined in the $I - pasting$, whenever they are well-defined on L and B. Indeed if $c \in L' \cap [c]$ and $f \in L' \cap [f]$, with $c \leq f$ holds for both $L' = L$ and $L' = B$, ϕ'_L preserving order, it must be either $c = 0_{L'}$ or $f = 1_{L'}$.

On the other hand, since L' is an OMP, then $c \leq f$ implies that $f = c \vee (f \wedge c')$, hence $[f] = [c] \vee ([f] \wedge [c]')$. $\qquad\square$

In the following, $L|I|B$ will refer to this particular construction, and L will be assumed to be stable. Furthermore, we will suppose that $\phi_L(y) = x$ is a free atom, and show that $L|I|B$ is stable whenever L is.

We start defining $J'(L) = (Q'_L, E'_L, T'_L)$ in the following way:

$$Q'_L = S_y \cup \{s \cup \{v_i\} | s \in S_{y'}, \phi_B(y) \neq v_i \in \mathcal{A}(B)\},$$
$$E'_L = \{[s, s'] | s, s' \in Q'_L, s \neq s'\},$$
$$T'_L = \{(s, [s, s'], s') | s, s' \in Q'_L, [s, s'] \in E'_L, s \neq s'\}.$$

Lemma 1. *$J'(L)$ is isomorphic to $J(L|I|B)$. Furthermore, for every $v_i \in \mathcal{A}(B)$ such that $\phi_B(y) \neq v_i$, the subgraph of $J'(L)$ induced by $S(y) \cup S(v_i)$ is isomorphic to $J(L)$.*

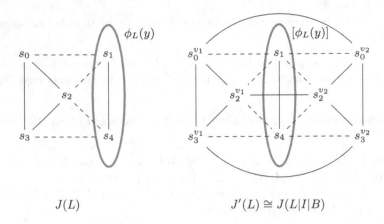

$$J(L) \qquad\qquad J'(L) \cong J(L|I|B)$$

Fig. 6. Construction of $J'(L)$, where L is the OMP of Fig. 2, I is as in Theorem 1, and B is a Boolean algebra with three atoms: $\mathcal{A}(B) = \{\phi_B(y), v_1, v_2\}$. Lines represent transitions in both directions. Dashed lines have an incidence with respect to $\phi_L(y)$ or $[\phi_L(y)]$, whereas solid lines are independent from them.

Proof. Every state $q \in Q'_L$ contains one, and only one, atom of B and since $\phi_L(y)$ is a free atom, it has one, and only one, atom for every Boolean algebra of L. Hence the elements of Q'_L coincide with the elements of $S(L|I|B)$. Since the construction of $J'(L)$ is completely determined by the set of states as in the construction of $J(L|I|B)$, the two transition systems $J(L|I|B)$ and $J'(L)$ are isomorphic (Fig. 6).

We also observe that for every atom $v_i \in \mathcal{A}(B)$ different from $\phi_B(y)$ the elements in $S(y) \cup S(v_i)$ coincide with the elements of $S(y) \cup S(y')$, which is the set of states of $J(L)$. From this observation it is easy to see that there is an isomorphism between $J(L)$ and any subgraph of $J'(L)$ induced by a set of states in the form $S(y) \cup S(v_i)$. □

This last lemma will permit to consider $J'(L)$ instead of $J(L|I|B)$.

Lemma 2. *If a region $H \in \mathcal{R}(J'(L))$ contains a state $s \cup \{v_i\} \in Q'_L$ and $S_{v_i} \not\subseteq H$ then $\forall s \cup \{v_j\} \in Q'_L : s \cup \{v_j\} \in H$.*

Proof. Since $S_{v_i} \not\subseteq H$ there must be a state $s' \cup \{v_i\} \notin H$, which means that the event $[s, s']$ is an event labeling a transition exiting H. Now suppose that there is a state $s \cup \{v_j\} \in Q'_L$ that doesn't belong to H. This means that the transition from $s \cup \{v_j\}$ to $s' \cup \{v_j\} \in Q'_L$ does not exit H, but such a transition is also labeled $[s, s']$, which is not possible since that would mean that H is not a region. □

Lemma 3. *Every atomic region of $J'(L)$ is in the form S_x for $x \in \mathcal{A}(L|I|B)$.*

Proof. Assume that there is an atomic region $H \notin \{S_x \mid x \in \mathcal{A}(L|I|B)\}$. Consider the subgraphs induced by $S_y \cup S_{v_i}$ for all the $\phi_B(y) \neq v_i \in B$ and call

$H_{v_i} \subset H$ the sets $H \cap (S_y \cup S_{v_i})$. All the H_{v_i} are atomic regions in every subgraph of $J'(L)$ induced by $S_y \cup S_{v_i}$, since they are all isomorphic to $J(L)$. The regions H_{v_i} can be atomic or not. If they are atomic, then they must coincide with an atomic region S_x, with $y \neq x \in L_n$. Hence, after Lemma 2, $H \in \{S_x\}_{x \in L_{n+1}}$. If they are not atomic, then there are $H'_{v_1} \subset H_{v_1}, ..., H'_{v_k} \subset H_{v_k}$ from which we can make the region $\bigcup_{i \in \{1,...,k\}} H'_{v_i} \subset H$, hence H is not atomic. $\qquad\square$

Theorem 2. *Let L be a stable* OMP*, and $x \in \mathcal{A}(L)$ be a free atom. Let B be a finite Boolean algebra, and $a \in \mathcal{A}(B)$. Let $I = \{0, 1, y, y'\}$, and define $\phi_L : I \to L$, and $\phi_B : I \to B$ such that $\phi_L(y) = x$, and $\phi_B(y) = a$. Then $L|I|B$ induced by the V-formation $(I, L, B, \phi_L, \phi_B)$ is stable.*

Proof. We wish to show that $H(J(L|I|B)) \simeq L|I|B$. With Lemma 1, we reduce it to showing that $H(J'(L)) \simeq L|I|B$. Since $H(J'(L))$ is a finite OMP, it is characterised by the orthogonality relation among its atoms. Now, Lemma 3 states that each atom of $H(J'(L))$ corresponds to an atom of $L|I|B$, and it was shown in [6] that every atom of $L|I|B$ must be an atom of $H(J'(L))$.

For each pair of elements $x \in \mathcal{A}(L)$, $y \in \mathcal{A}(B)$, there is a state in $s \in J(L|I|B)$ such that $[x] \in s$ and $[y] \in s$. Hence, the pasting must preserve incompatibility. Since the pasting also preserves orthogonality, we have that $L|I|B$, and $H(J(L|I|B))$ have same collection of atoms, with identical orthogonality relations. As it was shown in [7], this is sufficient to state that $H(J(L|I|B)) \simeq L|I|B$.

$\qquad\square$

3.3 Stability of Atom Refinement

The operation of refining an atom of an OMP into two new atoms preserves stability.

Theorem 3. *Let L be a stable* OMP*. Let $x \in \mathcal{A}(L)$. Consider $M_a = (\mathcal{A}(L) \setminus \{x\}) \cup \{y, z\}$, in which all orthogonal atoms of L remain orthogonal in M_a, and all atoms orthogonal to x in L are orthogonal to both y and z in M_a. Then the* OMP *M generated by M_a is stable.*

Proof. We will consider only the atoms of L and M, and states as represented by maximal cliques of $\$$ as in Proposition 1. Let $S_{x'}$ be the set of states of L not containing x. By construction of M_a, for each state $s \in S_x$ of L there are two states of M, in S_y and S_z respectively. Furthermore, the states of $S_{x'}$ all contain an atom orthogonal to x in L, and it will be orthogonal to both y and z in M. Thus, $S_{x'}, S_y$ and S_z constitute a partition of the states of M.

Starting from the states of M as partitioned above, it is possible to define the following sets of events: $F_{x'} = \{[s, s'] \mid s, s' \subset S_{x'}, s \neq s'\}$, $E_{y,z} = \{[s, s'] \mid s \in S_y, s' \in S_z\}$, $E_y = \{[s, s'] \mid s, s' \in S_y\}$, $E_z = \{[s, s'] \mid s, s' \in S_z\}$, $E_{x',y} = \{[s, s'] \mid s \in S_{x'}, s' \in S_y\}$ and $E_{x',z} = \{[s, s'] \mid s \in S_{x'}, s' \in S_z\}$.

Let A_y be the transition system with the following sets of states and events: $S_{x'} \cup S_y$ and $E_{x'} \cup E_y \cup E_{x',y}$, let, symmetrically, A_z be the transition system whose states are $S_{x'} \cup S_z$ and whose events are $E_{x'} \cup E_z \cup E_{x',z}$. We note that both $\mathcal{R}(A_y)$ and $\mathcal{R}(A_z)$ are isomorphic to the regions of the saturated system $J(L)$ since in both cases of $\mathcal{R}(A_y)$ and $\mathcal{R}(A_z)$, atoms y and z replace uniformly x. Moreover, since states S_y and S_z are disjoint, it is possible to construct the CETS $A = A_y \cup A_z$ endowed by all the new events in $E_{y,z}$. We note that $\mathcal{R}(A)$ must contain $\mathcal{R}(J(M))$ since CETS A, having less events than $J(M)$, can have less constraints in the construction of its regions. We want to show that $\mathcal{R}(A) = \mathcal{R}(J(M))$. Let, by contradiction, r be a region in $\mathcal{R}(A)$ not belonging to $\mathcal{R}(J(M))$.

If $r \subseteq S_{x'}$, then $r \in \mathcal{R}(J(M))$ since all the labels in $E_{x'}$ belong to both CETS A and $J(M)$ and the new events in $E_{x',y}$ and $E_{x',z}$ are distinct copies of the original events $E_{x',x}$ in $J(L)$, so they do not create new regions. If $r \subseteq S_y$ and, symmetrically, for $r \subseteq S_z$ then r must be a region in $\mathcal{R}(J(M))$ since all the labels in E_y and $E_{x',y}$, and symmetrically E_z and $E_{x',z}$ are, by construction, isomorphic to the labels $E_{x',x}$ in $J(L)$ and all the new labels $E_{y,z}$ are exiting from or, respectively, entering in r. The only remaining case could be for r being a minimal region in $\mathcal{R}(A)$ and a non minimal region in $\mathcal{R}(J(M))$ but this would be in contradiction with y and z being atoms in M.   □

4   Applying the Results to Prove Stability

It was shown, in [6], that some classes of OMPs are stable.

The first class regards the degenerate case in which the OMP is a Boolean algebra. Indeed, the result is here trivial, since all transitions of the saturated transition system carry different labels. Systems having a Boolean algebra for structure of regions are characterised by being fully sequential, the whole system is bound to a single locality. An example of this class is on the left in Fig. 7.

The second class, is that of OMPs consisting of a collection of Boolean algebras such that their pairwise intersections all coincide in the same Boolean subalgebra. This case has to be interpreted as the class of systems the sequential components of which, all synchronise through the same channel. One can think of several systems, pairwise independent but for a resource shared by all of them (see Fig. 7 for an example).

Finally, it was shown that the $\{0, 1\}$-pasting of two stable OMPs is stable. The $\{0, 1\}$-pasting of two OMPs corresponds to their disjoint union, but for identification of the maximal and minimal elements as in Fig. 8. This composition operation corresponds to the parallel composition of the corresponding operand systems. Indeed, the two systems are simply considered as a whole, although they remain independent, they do not synchronise or exchange information.   This operation allows to compose stable OMPs, with the certainty that the compound will be stable. The only requirement is that the operands are stable, so one can compose any two OMP from the aforementioned classes. Furthermore, the result of the composition being stable, it can be itself an operand for a further composition, and so this composition can be iterated, in order to generate a wide class of stable systems. As an operation, it is associative and commutative. However, this kind of composition is very restrictive, since it does not allow to specify but

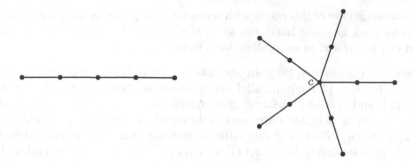

Fig. 7. To the left, a Boolean algebra with 5 atoms. To the right, an OMP consisting of 5 Boolean algebras of 3 atoms each, all sharing exactly the same atom c, hence all other atoms are free. Both OMPs are represented as *Greechie diagrams*, and both are stable.

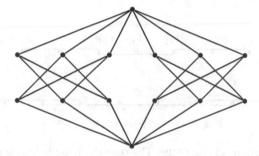

Fig. 8. *Hasse diagram* of the parallel composition of two Boolean algebras with 3 atoms each. It is obtained from the V-formation in Fig. 4, when replacing I by **2**, and morphisms accordingly.

an empty interface for the components. This issue has been tackled in the present work: in the last section we have presented a composition operation, which preserves stability of its operands, but allows to specify an interface. An interface is the identification of two local states, one in each of the operands, that will serve as a communication channel, allowing them to synchronise their behaviours. The result concerns the extension of a system with one single sequential component. Since the OMP resulting from such a composition is again stable, it can be taken as operand for further composition, and in this fashion, the composition can be iterated.

The main limitation of the composition operation we have presented is that it only allows for interfaces consisting of a single local state, and it would be suitable to extend the result so as to allow for larger interfaces, that could permit to model more complex communication protocols. In this sense, the solution we provide is a refinement operation, that preserves stability.

The intended use of this result with respect to the previous composition operation is twofold. One one hand, the atom serving as interface in the compound system can be refined so as to allow for a richer interface.

Example 9. Consider two Boolean algebras B_1, and B_2 with three atoms each, $\mathcal{A}(B_i) = \{a_i, b_i, c_i\}$. Their parallel composition is shown in Fig. 8. Let $I = \{0, x, x', 1\}$ and consider the two OMP-morphisms $\phi_i : I \to B_i$, such that $\phi_i(x) = c_i$. B_1 is a stable OMP, and c_1 is clearly a free atom, so after Theorem 2, $L = B_1|I|B_2$ is a stable OMP. L is isomorphic to the OMP depicted in Fig. 2, by considering $b_3 = [\phi_1(x)] = [\phi_2(x)]$. Since L is stable, and b_5 is a free atom, we can compose it with a new Boolean algebra B_3, by means of the morphisms ϕ_L and ϕ_3, provided by $\phi_L(x) = b_5$ and $\phi_3(x) = c_3$. The Greechie diagram of $L|I|B_3$ is depicted at the top of Fig. 9. Thanks to Theorem 3, we can now split any atom of $L|I|B_3$, obtaining, for example, the stable OMP depicted at the bottom of Fig. 9.

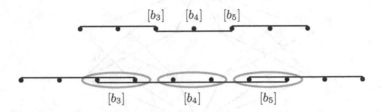

Fig. 9. Greechie diagrams of two OMPs. The OMP depicted below, is obtained from the one depicted above, by refining the atoms as shown. Since the OMP above is stable, so is the one below.

On the other hand, a free atom of the operands can be refined prior to the operation. In this way, only one of the two refining atoms will be used as interface with the appended sequential component, the other one remaining free for further composition. With this method, one can iterate the composition operation without worrying about exhaustion of available free atoms of the original system.

Example 10. Consider a system made of two sequential components each of which can get to a state for which they require the same resource. If each of these components can be in two additional states, the regional OMP for this system is represented as L_1 in Fig. 10. B_1 and B_2 represent the two sequential components, and x_1, x_2 correspond to their mutually exclusive states. B_c represents the state of the resource c, it can be in state x_i, indicating that B_i holds c, or in state y_1, indicating that c is available. When c is available, no other component is involved in the coresponding local state of B_c, and so y_1 is a free atom. L_1 is isomorphic to the OMP at the top of Fig. 9, and was shown to be stable in Example 9.

We may want to make the resource c available for a third sequential component B_3, so we can use Theorem 2 to compose L_1 and B_3 on y_1, obtaining the stable $L_2 = L_1|I|B_3$ a shown in Fig. 10. However, in this new compound system, the resource must be held by one of the three components B_i, as B_c has no more free atoms. No additional component can be added to the system, to compete for c. Instead of performing the composition $L_1|I|B_3$ directly, we can first make use of Theorem 3, and refine y_1 in L_1 into two new free atoms x_3, y_2, thus obtaining the OMP L_3 of Fig. 10. We can now compose it with B_3 on x_3, thus obtaining $L_4 = L_3|I|B_3$, which is stable. One can see that y_2 remains a free atom, a local state representing that the resource is available.

Analogously, y_2 can be further refined into x_4, y_3, so as to add a sequential component B_4 on x_4, to compete for the resource. This process can be iterated, to obtain a system with n sequential components competing for the same resource. Its regional OMP will then be L_5, as in Fig. 10. Note that for each B_i any atom but x_i is free, and so it can be refined to increase the number of local states of each sequential component.

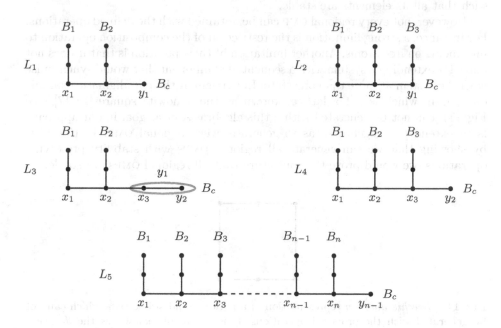

Fig. 10. Support for Example 10: refining a free atom, and iterating composition

5 Conclusion

The results we have presented in this work build upon previous results, and in many aspects, their proofs rely on them.

For instance, regional OMPs were studied for the first time in [4]. In that contribution, *primeness* was a central property of regional OMPs where an OMP is *prime* if, for every pair of distinct elements, there exists a state that contains one element and not the other. In [1], the equivalence of richness and primeness for OMPs was formally proved and this allowed to us to exploit the formal properties of rich OMPs in this contribution.

On the other hand, the analysis that was performed in [7] regarding suffi-ciency of atoms for the construction of the saturated transition system, allowed us to importantly simplify the proofs we have presented here.

Finally, in [6], stability of OMPs was shown for particular subclasses. Further-more, the parallel composition operation was shown to preserve stability, setting the ground for a different approach to prove the stability conjecture.

Indeed, in the present work, we have formalised two additional operations which preserve stability. One corresponds to the refinement of a local state into two. The second corresponds to the extension of a system with a sequential component which synchronises with it over a local state which is not already a synchronisation. With these elements, we can define an algebra of system OMPs, such that all its elements are stable.

However, not every regional OMP can be obtained with the defined operations. For instance, a strong limitation is the restriction of the composition operation to interfacing on free atoms. Another limitation of this operation is that it does not allow for extending a system with a sequential component that would synchronise with the system at more than one state. In particular, the paradigmatic example of an OMP which is not a lattice, known as the Janowitz square [14,17] (see Fig. 11), can not be generated within this algebra. A clear goal in our approach is to extend this algebra so as to generate every regional OMP. In this way, by showing that we can generate all regional OMPs with stability preserving operations, we would prove the conjecture that all regional OMPs are stable.

Fig. 11. *Greechie diagram* representation of a regional and stable OMP which can not be generated with the proposed operations. It is commonly known as the *Janowitz square*.

Note that the conjecture was first posed in a categorical setting. Indeed, J and H form a pair of adjoint functors between the category of (prime and coherent)[6] OMPs, and the category of CETS (with suitably defined morphisms). Unsurprisingly, this is the same kind of relation we find between the category

---

[6] In the terminology of [4].

Petri net systems, and the category of labelled transition systems (for example, in the elementary case). Indeed, the case graph construction, and the synthesis of saturated net system are seen as functors from one category to the other, and these are adjoint to each other. In this elementary framework however, it is shown that this adjunction is in fact a *co-reflection* [15], which implies that iterating alternated case graph, and synthesis constructions does only yield isomorphic elements in each of the two categories. Thus, the conjecture posed in [4], can be restated as emulating this result: (H, J) form a co-reflection.

Acknowledgements. We wish to thank Lucia Pomello, and Luca Bernardinello for the fruitful discussions. We are also grateful to the anonymous reviewers for their useful comments.

References

1. Adobbati, F., Ferigato, C., Gandelli, S., Aubel, A.P.: Two operations for stable structures of elementary regions. In: van der Aalst, W.M.P., Bergenthum, R., Carmona, J. (eds.) Proceedings of the International Workshop ATAED 2019 Satellite event of the conferences: ICATPN 2019 and ACSD 2019, CEUR Workshop Proceedings, Aachen, Germany, 25 June 2019, vol. 2371, pp. 36–53. CEUR-WS.org (2019)
2. Badouel, E., Bernardinello, L., Darondeau, P.: Petri Net Synthesis. TTCSAES. Springer, Heidelberg (2015). https://doi.org/10.1007/978-3-662-47967-4
3. Bernardinello, L.: Synthesis of net systems. In: Ajmone Marsan, M. (ed.) ICATPN 1993. LNCS, vol. 691, pp. 89–105. Springer, Heidelberg (1993). https://doi.org/10.1007/3-540-56863-8_42
4. Bernardinello, L., Ferigato, C., Pomello, L.: An algebraic model of observable properties in distributed systems. Theor. Comput. Sci. **290**(1), 637–668 (2003)
5. Bernardinello, L., Ferigato, C., Pomello, L., Aubel, A.P.: Synthesis of transition systems from quantum logics. Fundamenta Informaticae **154**(1–4), 25–36 (2017)
6. Bernardinello, L., Ferigato, C., Pomello, L., Aubel, A.P.: On stability of regional orthomodular posets. Trans. Petri Nets Other Models Concurrency **13**, 52–72 (2018)
7. Bernardinello, L., Ferigato, C., Pomello, L., Aubel, A.P.: On the decomposition of regional events in elementary systems. In: van der Aalst, W.M.P., Bergenthum, R., Carmona, J. (eds.) Proceedings of the International Workshop ATAED 2018 Satellite event of the conferences: ICATPN 2018 and ACSD 2018, CEUR Workshop Proceedings, Bratislava, Slovakia, 25 June 2018, vol. 2115, pp. 39–55. CEUR-WS.org (2018)
8. Birkhoff, G., Von Neumann, J.: The logic of quantum mechanics. Ann. Math **37**(4), 823–843 (1936)
9. Bruns, G., Harding, J.: Amalgamation of ortholattices. Order **14**(3), 193–209 (1997)
10. Clarke Jr, E.M., Grumberg, O., Kroening, D., Peled, D., Veith, H.: Model Checking. MIT Press, Cambridge (2001)
11. Ehrenfeucht, A., Rozenberg, G.: Partial (set) 2-structures: part I: basic notions and the representation problem. Acta Informatica **27**(4), 315–342 (1990)
12. Ehrenfeucht, A., Rozenberg, G.: Partial (set) 2-structures: part II: state spaces of concurrent systems. Acta Informatica **27**(4), 343–368 (1990)

13. Foulis, D.J., Randall, C.H.: Operational statistics. I: basic concepts. J. Math. Phys. **13**(11), 1667–1675 (1972)
14. Hughes, R.I.G.: The Structure and Interpretation of Quantum Mechanics. Harvard University Press, Cambridge (1989)
15. Nielsen, M., Rozenberg, G., Thiagarajan, P.S.: Elementary transition systems. Theor. Comput. Sci. **96**(1), 3–33 (1992)
16. Nielsen, M., Rozenberg, G., Thiagarajan, P.S.: Transition systems, event structures and unfoldings. Inf. Comput. **118**(2), 191–207 (1995)
17. Pavel, P., Sylvia, P.: Orthomodular Structures as Quantum Logics. Kluwer Academic Publishers, Amsterdam (1991)
18. Petri, C.A.: General net theory: computing system design. In: Joint IBM-University of Newcastle upon Tyne Seminar, September 1976, Proceedings (1977)
19. Randall, C.H., Foulis, D.J.: Operational statistics. II: manuals of operations and their logics. J. Math. Phys. **14**(10), 1472–1480 (1973)
20. Rival, I.: The diagram. In: Reidel, D. (ed.) Graphs and Order, Series C: Mathematical and Physical Sciences, pp. 103–133 (1985)

Efficient Synthesis of Weighted Marked Graphs with Circular Reachability Graph, and Beyond

Raymond Devillers[1], Evgeny Erofeev[2], and Thomas Hujsa[3(✉)]

[1] Département d'Informatique, Université Libre de Bruxelles, 1050 Brussels, Belgium
rdevil@ulb.ac.be
[2] Department of Computing Science, Carl von Ossietzky Universität Oldenburg,
26111 Oldenburg, Germany
evgeny.erofeev@informatik.uni-oldenburg.de
[3] LAAS-CNRS, Université de Toulouse, CNRS, Toulouse, France
thujsa@laas.fr

Abstract. In previous studies, several methods have been developed to synthesise Petri nets from labelled transition systems (LTS), often with structural constraints on the net and on the LTS. In this paper, we focus on Weighted Marked Graphs (WMGs) and Choice-Free (CF) Petri nets, two weighted subclasses of nets in which each place has at most one output; WMGs have the additional constraint that each place has at most one input.

We provide new conditions for checking the existence of a WMG whose reachability graph is isomorphic to a given circular LTS, i.e. forming a single cycle; we develop two new polynomial-time synthesis algorithms dedicated to these constraints: the first one is LTS-based (classical synthesis) while the second one is vector-based (weak synthesis) and more efficient in general. We show that our conditions also apply to CF synthesis in the case of three-letter alphabets, and we discuss the difficulties in extending them to CF synthesis over arbitrary alphabets.

Keywords: Weighted Petri net · Weighted marked graph · Choice-free net · Synthesis · Weak synthesis · Labelled transition system · Cycle · Cyclic word · Circular solvability · Polynomial-time algorithm · P-vector · T-vector · Parikh vector

1 Introduction

Petri nets form a highly expressive and intuitive operational model of discrete event systems, capturing the mechanisms of synchronisation, conflict and concurrency. Many of their fundamental behavioural properties are decidable, allowing to model and analyse numerous artificial and natural systems. However, most

E. Erofeev—Supported by DFG through grant Be 1267/16-1 ASYST.
T. Hujsa—Supported by the STAE foundation/project DAEDALUS, Toulouse, France.

© Springer-Verlag GmbH Germany, part of Springer Nature 2021
M. Koutny et al. (Eds.): ToPNoC XV, LNCS 12530, pp. 75–100, 2021.
https://doi.org/10.1007/978-3-662-63079-2_4

interesting model checking problems are worst-case intractable, and the efficiency of synthesis algorithms varies widely depending on the constraints imposed on the desired solution. In this study, we focus on the Petri net synthesis problem from a labelled transition system (LTS), which consists in determining the existence of a Petri net whose reachability graph is isomorphic to the given LTS, and building such a Petri net solution when it exists.

In previous studies on analysis or synthesis, structural restrictions on nets encompassed *plain* nets (each weight equals 1; also called ordinary nets) [33], *homogeneous* nets (for each place p, all the output weights of p are equal) [29, 37], *free-choice* nets (the net is plain, and any two transitions sharing an input have the same set of inputs) [15,37], *join-free* nets (each transition has at most one input place) [14,28,29,37]. Recently, another kind of restriction has been considered, limiting the number of distinct labels of the LTS [5,6,24,25].

Depending on the constraints on the solution to be constructed, the complexity of the synthesis problem can vary widely: the problem can be solved in polynomial-time for bounded Petri nets [3], while aiming at *elementary net systems*, or at various other Petri net subclasses with fixed marking bound, makes the problem NP-complete [4,38].

In this paper, we study the solvability of LTS with weighted marked graphs (WMGs; each place has at most one output and one input) and choice-free nets (CF; each place has at most one output). Both classes are important for real-world applications, and are widely studied in the literature [10–12,19,21,27,35,36]. We focus mainly on finite *circular LTS*, i.e. strongly connected LTS that contain a unique *cycle*[1]. In this context, we investigate the *cyclic solvability* of a word w, meaning the existence of a Petri net solution to the finite circular LTS induced by the infinite *cyclic word* w^∞. These restrictions appear in practical situations, since various complex applications can be decomposed into subsystems satisfying such constraints [5,7,11,17,18,23,24,26].

Contributions. We study further the links between simple LTS structures and the reachability graph of WMGs and CF nets, as follows. First, we show that a binary (i.e. over a two-letter alphabet) LTS is CF-solvable if and only if it is WMG-solvable. Then, we develop new conditions for WMG-solving a cyclic word over an arbitrary alphabet, with a polynomial-time synthesis algorithm.

We show that a word over a three-letter alphabet is cyclically WMG-solvable iff it is cyclically CF-solvable, and that this result does not hold with four-letter alphabets. More generally, we discuss the difficulties of extending these results to CF synthesis over arbitrary alphabets.

We introduce the notion of *weak synthesis*, which aims at synthesising a Petri net from a given transition-vector Υ instead of a sequence: the solution obtained enables some sequence whose Parikh vector equals Υ. This allows to

---

[1] A set A of k arcs in a LTS G defines a cycle of G if the elements of A can be ordered as a sequence $a_1 \ldots a_k$ such that, for each $i \in \{1, \ldots, k\}$, $a_i = (n_i, \ell_i, n_{i+1})$ and $n_{k+1} = n_1$, i.e. the i-th arc a_i goes from node n_i to node n_{i+1} until the first node n_1 is reached, closing the path.

be less restrictive on the solution design. Then, we provide a polynomial-time algorithm for the weak synthesis of WMGs with circular reachability graphs.

Finally, we show that our weak synthesis algorithm performs generally much faster than the sequence-based algorithm.

Comparing with [20], we provide more details, we add the equivalence result on CF nets for three-letter alphabets in Subsection 4.4 and the new Sect. 5 on weak synthesis, with a new synthesis algorithm and the study of its complexity.

Organisation of the Paper. After recalling classical definitions, notations and properties in Sect. 2, we present the equivalence of CF- and WMG-solvability for 2-letter words in Sect. 3.

In Sect. 4, we focus on circular LTS: we give a new characterisation of WMG-solvability and a dedicated polynomial-time synthesis algorithm. We prove the equivalence between cyclic WMG and CF synthesis for three-letter alphabets. We also provide a number of examples showing that some of our results cannot be applied to the class of CF-nets over arbitrary alphabets.

Section 5 contains our study of the weak synthesis problem for WMGs with a circular reachability graph, with a new polynomial-time synthesis algorithm. Finally, Sect. 6 presents our conclusions and perspectives.

2 Classical Definitions, Notations and Properties

LTS, Sequences and Reachability. A *labelled transition system with initial state, LTS* for short, is a quadruple $TS = (S, \rightarrow, T, \iota)$ where S is the set of *states*, T is the (finite) set of *labels*, $\rightarrow \subseteq (S \times T \times S)$ is the *transition relation*, and $\iota \in S$ is the *initial state*. A label t is *enabled* at $s \in S$, written $s[t\rangle$, if $\exists s' \in S \colon (s, t, s') \in \rightarrow$, in which case s' is said to be *reachable* from s by the firing of t, and we write $s[t\rangle s'$. Generalising to any (firing) sequences $\sigma \in T^*$, $s[\varepsilon\rangle$ and $s[\varepsilon\rangle s$ are always true, with ε being the empty sequence; and $s[\sigma t\rangle s'$, i.e., σt is *enabled* from state s and leads to s' if there is some s'' with $s[\sigma\rangle s''$ and $s''[t\rangle s'$. For clarity, in case of long formulas we write $\lfloor_r \sigma \lfloor_s \tau \lfloor_q$ instead of $r[\sigma\rangle s[\tau\rangle q$, thus fixing some intermediate states along a firing sequence. A state s' is *reachable* from state s if $\exists \sigma \in T^* \colon s[\sigma\rangle s'$. The set of states reachable from s is noted $[s\rangle$.

Petri Nets and Reachability Graphs. A (finite, place-transition) *weighted Petri net*, or *weighted net*, is a tuple $N = (P, T, W)$ where P is a finite set of *places*, T is a finite set of *transitions*, with $P \cap T = \emptyset$ and W is a *weight* function $W \colon ((P \times T) \cup (T \times P)) \rightarrow \mathbb{N}$ giving the weight of each arc. A *Petri net system*, or *system*, is a tuple $\mathcal{S} = (N, M_0)$ where N is a net and M_0 is the *initial marking*, which is a mapping $M_0 \colon P \rightarrow \mathbb{N}$ (hence a member of $\mathbb{N}^P$) indicating the initial number of *tokens* in each place. The *incidence matrix* I of the net is the integer $P \times T$-matrix with components $I(p, t) = W(t, p) - W(p, t)$.

A place $p \in P$ is *enabled by* a marking M if $M(p) \geq W(p, t)$ for every transition $t \in T$. A transition $t \in T$ is *enabled by* a marking M, denoted by $M[t\rangle$, if for all places $p \in P$, $M(p) \geq W(p,t)$. If t is enabled at M, then t can *occur* (or *fire*) in M, leading to the marking M' defined by $M'(p) = M(p) - W(p,t) + W(t,p)$, denoted $M[t\rangle M'$. A marking M' is *reachable* from M if there is a sequence of firings leading from M to M'. The set of markings reachable from M is denoted by $[M\rangle$. The *reachability graph of* S is the labelled transition system $RG(S)$ with the set of vertices $[M_0\rangle$, the set of labels T, initial state M_0 and transitions $\{(M, t, M') \mid M, M' \in [M_0\rangle \wedge M[t\rangle M'\}$. A system S is *bounded* if $RG(S)$ is finite.

Vectors. The *support* of a vector is the set of the indices of its non-null components. Consider any net $N = (P, T, W)$ with its incidence matrix I. A *T-vector* (respectively *P-vector*) is an element of $\mathbb{N}^T$ (respectively $\mathbb{N}^P$); it is called *prime* if the greatest common divisor of its components is one (i.e., it is non-null and its components do not have a common non-unit factor). A *T-semiflow* ν of the net is a non-null T-vector such that $I \cdot \nu = 0$. A T-semiflow is called *minimal* when it is prime and its support is not a proper superset of the support of any other T-semiflow [36].

The *Parikh vector* $\mathbf{P}(\sigma)$ of a finite transition sequence σ is a T-vector counting the number of occurrences of each transition in σ, and the *support* of σ is the support of its Parikh vector, i.e., $supp(\sigma) = supp(\mathbf{P}(\sigma)) = \{t \in T \mid \mathbf{P}(\sigma)(t) > 0\}$.

Strong Connectedness and Cycles in LTS. The LTS is said *reversible* if, $\forall s \in [\iota\rangle$, we have $\iota \in [s\rangle$, i.e., it is always possible to go back to the initial state; reversibility implies the strong connectedness of the LTS.

A sequence $s[\sigma\rangle s'$ is a *cycle*, or more precisely a *cycle at (or around) state* s, if $s = s'$. A non-empty cycle $s[\sigma\rangle s$ is called *small* if there is no non-empty cycle $s'[\sigma'\rangle s'$ in TS with $\mathbf{P}(\sigma') \lneq \mathbf{P}(\sigma)$ (the definition of Parikh vectors extends readily to sequences over the set of labels T of the LTS). A cycle $s[\sigma\rangle s$ is *prime* if $\mathbf{P}(\sigma)$ is prime. TS has the *prime cycle property* if each small cycle has a prime Parikh vector.

A *circular LTS* is a finite, strongly connected LTS that contains a unique cycle; hence, it has the shape of an oriented circle. The circular LTS *induced by* a word $w = w_1 \ldots w_k$ is defined as $s_0[w_1\rangle s_1[w_2\rangle s_2 \ldots [w_k\rangle s_0$ with initial state s_0. All notions defined for labelled transition systems apply to Petri nets through their reachability graphs.

Some Petri Net Subclasses. A net $N = (P, T, W)$ is *plain* if no arc weight exceeds 1; *pure* if $\forall p \in P$: $(p^\bullet \cap {}^\bullet p) = \emptyset$, where $p^\bullet = \{t \in T \mid W(p,t) > 0\}$ and ${}^\bullet p = \{t \in T \mid W(t,p) > 0\}$; *choice-free* (CF) [13,36] or place-output-nonbranching (ON) [7] if $\forall p \in P$: $|p^\bullet| \leq 1$; a *weighted marked graph* (WMG) if $|p^\bullet| \leq 1$ and $|{}^\bullet p| \leq 1$ for all places $p \in P$. The WMGs form a subclass of the CF nets and contain the weighted T-systems (WTSs) of [35], also known as *weighted event graphs* (WEGs) in [32], in which $\forall p \in P$, $|{}^\bullet p| = 1$ and $|p^\bullet| = 1$. Plain WEGs are also known as *marked graphs* [12] or *T-nets* [15].

Isomorphism and Solvability. Two LTS $TS_1 = (S_1, \rightarrow_1, T, s_{01})$ and $TS_2 = (S_2, \rightarrow_2, T, s_{02})$ are isomorphic if there is a bijection $\zeta \colon S_1 \rightarrow S_2$ with $\zeta(s_{01}) = s_{02}$ and $(s, t, s') \in \rightarrow_1 \Leftrightarrow (\zeta(s), t, \zeta(s')) \in \rightarrow_2$, for all $s, s' \in S_1$. If an LTS TS is isomorphic to $RG(\mathcal{S})$, where $\mathcal{S}$ is a net system, we say that $\mathcal{S}$ *solves* TS. Solving a word $w = \ell_1 \dots \ell_k$ amounts to solve the acyclic LTS defined by the single path $\iota[\ell_1\rangle s_1 \dots [\ell_k\rangle s_k$. A finite word w is *cyclically solvable* if the circular LTS induced by w is solvable. An LTS is WMG- (or CF-)solvable if a WMG (or a CF system) solves it.

Separation Problems. Let $TS = (S, \rightarrow, T, s_0)$ be a given labelled transition system. The theory of regions [2] characterises the solvability of an LTS through the solvability of a set of *separation problems*. In case the LTS is finite, we have to solve $\frac{1}{2} \cdot |S| \cdot (|S|-1)$ states separation problems and up to $|S| \cdot |T|$ event/state separation problems, as follows:

- A *region* of $(S, \rightarrow, T, s_0)$ is a triple $(\mathbb{R}, \mathbb{B}, \mathbb{F}) \in (\mathbb{N}^S, \mathbb{N}^T, \mathbb{N}^T)$ such that for all $(s, t, s') \in \rightarrow$, $\mathbb{R}(s) \geq \mathbb{B}(t)$ and $\mathbb{R}(s') = \mathbb{R}(s) - \mathbb{B}(t) + \mathbb{F}(t)$. A region models a place p, in the sense that $\mathbb{R}(s)$ models the token count of p at the marking corresponding to s, $\mathbb{B}(t)$ (for *backward*) models $W(p, t)$, and $\mathbb{F}(t)$ (for *forward*) models $W(t, p)$.
- A *states separation problem* (SSP for short) consists of a set of states $\{s, s'\}$ with $s \neq s'$. It is solved by a region $(\mathbb{R}, \mathbb{B}, \mathbb{F})$ when $\mathbb{R}(s) \neq \mathbb{R}(s')$, meaning the region allows to discriminate between s and s'.
- An *event/state separation problem* (ESSP for short) consists of a pair $(s, t) \in S \times T$ with $\neg s[t\rangle$. It is solved by a region $(\mathbb{R}, \mathbb{B}, \mathbb{F})$ when $\mathbb{R}(s) < \mathbb{B}(t)$, meaning the region allows to exclude a forbidden transition from some state.

In the rest of this paper, we interpret these two separation problems in terms of places of the hoped-for Petri net system as follows:

- For each SSP $\{s, s'\}$, $s \neq s'$, the two states s and s' must be distinguished by a place p such that $M_s(p) \neq M_{s'}(p)$, i.e. p has a different number of tokens in the markings corresponding to the two states.
- For each ESSP (s, t) with $\neg s[t\rangle$, there must exist a place p such that $M_s(p) < W(p, t)$ for the marking M_s corresponding to state s, where W refers to the arcs of the hoped-for Petri net system.

Notice that if the LTS is infinite, also the number of separation problems (of each kind) becomes infinite.

A synthesis procedure does not necessarily lead to a connected solution. However, the technique of decomposition into prime factors described in [16, 17] can always be applied first, so as to handle connected partial solutions and recombine them afterwards. Hence, in the sequel, we focus on connected nets, w.l.o.g. In the next section, we consider the CF synthesis problem with two distinct labels.

3 Reversible Binary CF Synthesis

In this section, we relate CF- to WMG-solvability for binary reversible LTS.

Lemma 1 (Pure CF-solvability)
If a reversible LTS has a CF solution, it has a pure CF solution.

Proof. Let $TS = (S, \rightarrow, T, \iota)$ be a reversible LTS. If $t \in T$ does not occur in $\rightarrow$, TS is solvable iff $TS' = (S, \rightarrow, T \setminus \{t\}, \iota)$ is solvable and a possible solution of TS is obtained by adding to any solution of TS' a transition t and a fresh place p, initially empty, with an arc from p to t (e.g. with weight 1), so that p is not a side condition[2]. We can thus assume that each label of T occurs in $\rightarrow$.

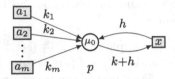

Fig. 1. A general pure ($h = 0$) or non-pure ($h > 0$) choice-free place p with initial marking μ_0. Place p has at most one outgoing transition named x. The set $\{a_1, \ldots, a_m\}$ comprises all other transitions, i.e., $T = \{x, a_1, \ldots, a_m\}$, and k_j denotes the weight of the arc from a_j to p (which could be zero).

The general form of a place in a CF solution is exhibited in Fig. 1. If $h = 0$, we are done, so that we shall assume $h > 0$. If $-h \leq k < 0$, the marking of p cannot decrease, and since x occurs in $\rightarrow$, the system cannot be reversible. If $k = 0$, for the same reason all the k_i's must be null too, $\mu_0 \geq h$, and we may drop p. Hence we assume that $k > 0$ and $\exists i : k_i > 0$.

Once x occurs, the marking of p is at least h, remains so, and since the system is reversible, all the reachable markings have at least h tokens in p. But then, if we replace p by a place p' with initially $\mu_0 - h$ tokens, the same k_i's and $h = 0$, we get exactly the same reachability graph, but with h tokens less in p' than in p. This will wipe out the side condition for p, and repeating this for each side condition, we get an equivalent pure and choice-free solution.   □

Theorem 1 (Reversible binary CF-solvability)
A binary reversible LTS is CF-solvable iff it is WMG-solvable.

Proof. If we have two labels, from Lemma 1, if there is a CF solution, there will be one with places of the form exhibited in Fig. 2, hence a WMG solution.   □

In the next section, the number of letters is no more restricted.

---

[2] A place p is a *side condition* if $^\bullet p \cap p^\bullet \neq \emptyset$.

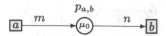

Fig. 2. A generic pure CF-place with two labels.

4 Cyclic WMG- and CF-Solvablity

In this section, we recall and extend conditions for WMG-solvability of some restricted classes of LTS formed by a single circuit, which were developed in [19].

We gradually study the separation problems – SSPs in Subsect. 4.1 and ESSPs in Subsect. 4.2 – for cyclic solvability with WMGs, leading to a language-theoretical characterisation of cyclically WMG-solvable sequences. The characterisation gives rise to a polynomial-time synthesis algorithm in Subsect. 4.3, which is shown to be more efficient than the classical synthesis approach.

Finally, in Subsect. 4.4, we study the extensibility of these results to the CF case: for three-letter alphabets, we show that a word is cyclically WMG-solvable iff it is cyclically CF-solvable; unfortunately, for arbitrary alphabets, we show with the help of examples that the other results cannot be directly extended.

In the following, two distinct labels a and b are called *(circularly) adjacent* in a word w if $w = (w_1 a b w_2)$ or $w = (b w_3 a)$ for some $w_1, w_2, w_3 \in T^*$. We denote by $p_{a,*}$ any place $p_{a,b}$ where b is adjacent to a. Also, if $T = \{t_0, t_1, \ldots, t_m\}$ with $m > 0$, at least one label is adjacent to t_0, and at each point at least one label is adjacent to the ones we distinguished so far, until we get the whole set T; we can thus start from any label t_i instead of t_0.

Theorem 2 (Sufficient condition for cyclic WMG-solvability [19])
Consider any word w over any finite alphabet T such that $\mathbf{P}(w)$ is prime. Suppose the following: $\forall u = w_{|t_1 t_2}$ (i.e., the projection[3] of w on $\{t_1, t_2\}$) for some distinct circularly adjacent labels t_1, t_2 in w, $u = v^\ell$ for some positive integer ℓ such that $\mathbf{P}(v)$ is prime, and v is cyclically solvable by a circuit (i.e., a circular Petri net system). Then, w is cyclically solvable with a WMG.

Theorem 3 (Cyclic WMG-solvability of ternary words [19])
Consider a ternary word w (with three letters in its alphabet T) with Parikh vector (x, x, y) such that $\gcd(x, y) = 1$. Then, w is cyclically solvable with a WMG if and only if, for any pair $t_1 \neq t_2 \in T$ such that $w = (w_1 t_1 t_2 w_2)$ or $w = (t_2 w_3 t_1)$, $u = v^\ell$ for some positive integer ℓ with $u = w_{|t_1 t_2}$, $\mathbf{P}(v)$ is prime, and v is cyclically solvable by a circuit.

For a circular LTS, the solvability of its binary projections by circuits is a sufficient condition, as specified by Theorem 2, but it turns out not to be

---

[3] The projection of a word $w \in A^*$ on a set $A' \subseteq A$ of labels, noted $w_{|A'}$, is the word obtained by erasing in w all the occurrences of labels belonging to $A \setminus A'$. For example, the projection of the word $w = \ell_1 \ell_2 \ell_3 \ell_2$ on the set $\{\ell_1, \ell_2\}$ is the word $\ell_1 \ell_2 \ell_2$.

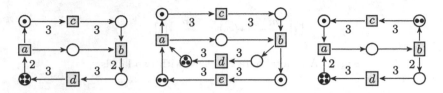

Fig. 3. The WMG on the left solves *aacbbdabd* cyclically, and the WMG in the middle solves *aacbbeabd* cyclically. On the right, the WMG solves *abcabdabd* cyclically.

a necessary one. Indeed, for the cyclically solvable sequence $w_1 = aacbbdabd$ (cf. left of Fig. 3), its binary projection on $\{a, b\}$ is $w_{1|a,b} = aabbab$ which is not cyclically solvable with a WMG (neither generally solvable). Looking only at the Parikh vector of the sequence is also not enough to establish its cyclic (un)solvability. For instance, sequences $w_2 = abcabdabd$ and $w_3 = abcbadabd$ are Parikh-equivalent: $\mathbf{P}(w_2) = \mathbf{P}(w_3) = (3, 3, 1, 2)$ (and also Parikh-equivalent to w_1), but w_2 is cyclically solvable with a WMG (e.g. with the WMG on the right of Fig. 3) and w_3 is not WMG-cyclically solvable.

All the binary projections of w_1 and w_3 are cyclically WMG-solvable, except $w_{i|a,b}$. Only the unsolvability of $w_{3|a,b}$ implies the unsolvability of w_3. Since all the w_i are Parikh-equivalent, then so are their binary projections. Thus, we have to analyse the sequences themselves, without abstracting to Parikh vectors. Since the projections $w_{1|a,b}$ and $w_{3|a,b}$ are equivalent (up to cyclic rotation and swapping a and b), it is not sufficient to check the 'problematic' binary projections. We then study the conditions for solvability of separation problems.

4.1   SSPs for Prime Cycles

For any word $w = t_0 \ldots t_k$, for $0 \le i, j \le k$ such that $i \ne j$, we note $\mathbf{P}_{ij} = \mathbf{P}(t_i t_{i+1} \ldots t_{j-1})$ if $i < j$ and $\mathbf{P}_{ij} = \mathbf{P}(t_i t_{i+1} \ldots t_{k-1} t_k t_0 t_1 \ldots t_{j-1})$ if $i > j$.

Lemma 2 (SSPs are solvable for prime cycles). *For a cyclic transition system $TS = (S, \to, T, s_0)$ defined by some word $w = t_0 \ldots t_k$, where $S = \{s_0, \ldots, s_k\}$, $\to = \{(s_{i-1}, t_{i-1}, s_i) \mid 1 \le i \le k\} \cup \{(s_k, t_k, s_0)\}$, if $\mathbf{P}(w)$ is prime then all the SSPs are solvable.*

Proof. If $|T| = 1$, then $k = 0$ and $|S| = 1$, so that there is no SSP to solve. We may thus assume $|T| > 1$.

For each pair of distinct labels $a, b \in T$ that are adjacent in TS, construct places $p_{a,b}$ (and $p_{b,a}$ since adjacency is commutative) as in Fig. 2 with

$$m = \frac{\mathbf{P}(w)(b)}{\gcd(\mathbf{P}(w)(a), \mathbf{P}(w)(b))}, \quad n = \frac{\mathbf{P}(w)(a)}{\gcd(\mathbf{P}(w)(a), \mathbf{P}(w)(b))}, \tag{1}$$

and $\mu_0 = n \cdot \mathbf{P}(w)(b)$. Clearly, the markings of $p_{a,b}$ reachable by repeatedly firing $u = w_{|ab}$ are always non-negative, and the initial marking is reproduced after each repetition of the sequence u. Consider two distinct states $s_i, s_j \in S$ (w.l.o.g.

$i < j$). We now demonstrate that there is at least one place of the form $p_{a,b}$ such that $M_i(p_{a,b}) \neq M_j(p_{a,b})$, where M_l denotes the marking corresponding to state s_l for $0 \leq l \leq k$. If $j - i = 1$, then any place of the form $p_{t_i,*}$ distinguishes states s_i and s_j. The same is true if $j - i > 1$ but $\forall l \in [i, j-1] : t_l = t_i$. Otherwise, choose some letter a from $t_i \ldots t_{j-1}$ and an adjacent letter b. Then $M_j(p_{a,b}) = M_i(p_{a,b}) + m \cdot \mathbf{P}_{ij}(a) - n \cdot \mathbf{P}_{ij}(b)$. If $M_i(p_{a,b}) \neq M_j(p_{a,b})$, place $p_{a,b}$ distinguishes s_i and s_j. Otherwise we have $m \cdot \mathbf{P}_{ij}(a) = n \cdot \mathbf{P}_{ij}(b)$, hence, due to the choice of m and n:

$$\frac{\mathbf{P}_{ij}(a)}{\mathbf{P}(w)(a)} = \frac{\mathbf{P}_{ij}(b)}{\mathbf{P}(w)(b)}$$

(so that b also belongs to $t_i \ldots t_{j-1}$). Consider some other letter c which is adjacent to a or b. If place $p_{a,c}$ distinguishes s_i and s_j, we are done. Otherwise, due to the choice of the arc weights for these places, we have

$$\frac{\mathbf{P}_{ij}(a)}{\mathbf{P}(w)(a)} = \frac{\mathbf{P}_{ij}(c)}{\mathbf{P}(w)(c)} = \frac{\mathbf{P}_{ij}(b)}{\mathbf{P}(w)(b)}.$$

Since $t_i \ldots t_{j-1}$ is finite, by progressing along the adjacency relation, either we find a place which has different markings at s_i and s_j, or for all $a, b \in supp(t_i \ldots t_{j-1})$ we have

$$\frac{\mathbf{P}_{ij}(a)}{\mathbf{P}(w)(a)} = \frac{\mathbf{P}_{ij}(b)}{\mathbf{P}(w)(b)}.$$

If $supp(t_i \ldots t_{j-1}) = supp(w)$, $\mathbf{P}(w)$ is proportional to $\mathbf{P}(t_i \ldots t_{j-1})$, but since $t_i \ldots t_{j-1}$ is smaller than w (otherwise $s_i = s_j$) this contradicts the primality of $\mathbf{P}(w)$. Hence, there exist adjacent c and d such that $c \in supp(w) \setminus supp(t_i \ldots t_{j-1})$ and $d \in supp(t_i \ldots t_{j-1})$. For the place $p_{c,d}$ we have $M_j(p_{c,d}) \neq M_i(p_{c,d})$. □

This property has some similarities with Theorem 4.1 in [22], but the preconditions are different. The reachability graph of any CF system, hence of any WMG, satisfies the prime cycle property [8,9]. Thus, primeness of a sequence avoids solving SSPs when aiming at these two classes of Petri nets.

4.2   ESSPs in Cyclic WMG-Solvability

Now, we develop further conditions for the cyclic WMG-solvability.

Lemma 3 (Special form of WMG solutions for cycles). *If $w \in T^*$ is cyclically solvable by a WMG, there exists a WMG $\mathcal{S} = ((P, T, W), M_0)$, where P consists of places $p_{a,b}$, for each pair of distinct circularly adjacent a and b (i.e., either $w = u_1 a b u_2$ or $w = b u_3 a$).*

Proof. Consider a sequence $w = t_0 \ldots t_k$, where $\mathbf{P}(w)$ is prime. Let us assume that the system $((P, T, W), M_0)$ is a WMG solving w cyclically. Due to the definition of WMGs, all the places that we have to consider are of the form schematised in Fig. 4. The arc weights may differ due to the parameter $l > 0$, but the ratio $\frac{W(a, p_{a,b})}{W(p_{a,b}, b)} = \frac{m}{n}$ is determined by the Parikh vector of w and its cyclic solvability; the initial marking is to be defined. Moreover, we have to consider only those places which are connected to the pairs of circularly adjacent transitions in w. Indeed, if $w = u_1 \lfloor_{s_i} a \lfloor_{s_{i+1}} b\, u_2$, where $b \neq a$, s_i is the state reached after performing u_1 and s_{i+1} is the state reached after performing $u_1 a$, then any place that solves the ESSP $\neg M_i[b\rangle$ is an input place for b. On the other hand, any place whose marking at M_{s_i} differs from its marking at $M_{s_{i+1}}$ is connected to a. Hence, a place $p \in P$ solving $\neg M_i[b\rangle$ is of the form $p_{a,b}$. Since p is only affected by a and b, it also disables b at all the states between s_l and s_i in w when it is of the form $w = u_3 \lfloor_{s_j} t_j \lfloor_{s_{j+1}} b^+ \lfloor_{s_l} u_4 \lfloor_{s_i} a b u_2$ with $\mathbf{P}(u_4)(b) = 0$ (in the case there is no b in the prefix between s_0 and abu_2, $s_l = s_0$). Analogously, if $t_j \neq b$, there must be a place $q \in P$ of the form $p_{t_j,b}$ that solves $\neg M_{s_j}[b\rangle$. Doing so, we ascertain that the places of the form schematised in Fig. 4 for the adjacent pairs of transitions are sufficient to handle all the ESSPs.

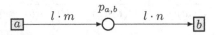

Fig. 4. A general place from a to b in a WMG solution of w: $m = \mathbf{P}(w)(b)$, $n = \mathbf{P}(w)(a)$, l may be any multiple of $1/\gcd(m,n)$.

In fact, for each pair of adjacent transitions a and b in w, a single place of the form $p_{a,b}$ is sufficient. Indeed, assume there are $p_1, p_2 \in P$ of the form $p_{a,b}$. If $\dfrac{M_0(p_1)}{\gcd(W(a,p_1),W(p_1,b))} \geq \dfrac{M_0(p_2)}{\gcd(W(a,p_2),W(p_2,b))}$ then for any $M \in [M_0\rangle$, $M(p_1) < W(p_1,b)$ implies $M(p_2) < W(p_2,b)$. Hence, p_1 is redundant in the system. It means we can choose l as we want (among the multiples of $1/\gcd(m,n)$) and only keep the place of the form $p_{a,b}$ in any solution with the smallest initial marking. Note that it may happen that we need a place $p_{a,b}$, but not $p_{b,a}$. $\square$

The existence of a WMG solution of this special form allows us to establish a necessary condition for the cyclic solvability of sequences.

Lemma 4 (A necessary condition for cyclic solvability with a WMG).
If $w \in T^*$ is cyclically solvable by a WMG, then for any adjacent transitions a and b in w, and any two occurrences of ab in $w = u_1 \lfloor_{s_r} ab \ldots \lfloor_{s_q} ab u_2$, the inequality

$$\frac{\mathbf{P}_{rj}(b) - 1}{\mathbf{P}_{rj}(a)} < \frac{m}{n} < \frac{\mathbf{P}_{jq}(b) + 1}{\mathbf{P}_{jq}(a)} \tag{2}$$

holds true, where m, n are as in (1), $r \leq j \leq q$, and the right inequality is omitted when $\mathbf{P}_{jq}(a) = 0$ and the left inequality is omitted when $\mathbf{P}_{rj}(a) = 0$.

Proof. Let w be cyclically solvable with a WMG $\mathcal{S} = ((P,T,W),M_0)$ as in Lemma 3, and place $p \in P$ be of the form $p_{a,b}$ (as in Fig. 4, with $l = 1$ and a well chosen initial marking) for an adjacent pair ab. Choose two ab's in $w = u_1 \lfloor_{s_r} a \lfloor_{s_{r+1}} b \lfloor_{s_{r+2}} \cdots \lfloor_{s_q} a \lfloor_{s_{q+1}} b\, u_2$ with possibly other letters between

s_{r+2} and s_q (if there is only one ab, apply the argumentation while wrapping around w circularly, i.e., $s_r[w\rangle s_q$). Since p solves ESSPs $\neg s_r[b\rangle$ and $\neg s_q[b\rangle$, the next inequalities hold true, where μ_r denotes the marking of $p_{a,b}$ at state s_r:

$$
\begin{aligned}
\neg s_r[b\rangle: \quad & \mu_r & < n \\
s_{r+1}[b\rangle: \quad & \mu_r + m & \geq n \\
\forall j: r \leq j \leq q: \quad & \mu_r + \mathbf{P}_{rj}(a) \cdot m - \mathbf{P}_{rj}(b) \cdot n & \geq 0 \\
\neg s_q[b\rangle: \quad & \mu_r + \mathbf{P}_{rq}(a) \cdot m - \mathbf{P}_{rq}(b) \cdot n & < n
\end{aligned}
\tag{3}
$$

From the first and the third line of (3) we get $\mathbf{P}_{rj}(a) \cdot m - \mathbf{P}_{rj}(b) \cdot n > -n$. This implies:

$$
\frac{\mathbf{P}_{rj}(b) - 1}{\mathbf{P}_{rj}(a)} < \frac{m}{n} \text{ when } r < j \leq q.
\tag{4}
$$

From the third and the fourth line of (3) we obtain

$$
(\mathbf{P}_{rq}(a) - \mathbf{P}_{rj}(a)) \cdot m - (\mathbf{P}_{rq}(b) - \mathbf{P}_{rj}(b)) \cdot n < n.
$$

If $\mathbf{P}_{jq}(a) \neq 0$, since $\mathbf{P}_{rq} = \mathbf{P}_{rj} + \mathbf{P}_{jq}$ this inequality can be written as

$$
\frac{m}{n} < \frac{\mathbf{P}_{jq}(b) + 1}{\mathbf{P}_{jq}(a)}.
\tag{5}
$$

Thus, from (4) and (5) we have a necessary condition for solvability.   □

In particular, Lemma 4 explains the cyclic unsolvability of the word $w_3 = \lfloor_{s_r} abcb \lfloor_{s_j} ad \lfloor_{s_q} abd$. Indeed, $\mathbf{P}(w_3)(b) = 3 = \mathbf{P}(w_3)(a)$, so that $m/n = 1$ and $1 \not< \frac{0+1}{1} = \frac{\mathbf{P}_{jq}(b)+1}{\mathbf{P}_{jq}(a)}$. Moreover, the necessary condition for cyclic sovability from Lemma 4 extends to a sufficient one in the following sense.

Lemma 5 (A sufficient condition for cyclic solvability by a WMG). *If $w \in T^*$ has a prime Parikh vector, and for each circularly adjacent ab pair in $w = \ldots \lfloor_{s_q} ab \ldots$, the inequality*

$$
\frac{m}{n} < \frac{\mathbf{P}_{jq}(b) + 1}{\mathbf{P}_{jq}(a)}
\tag{6}
$$

holds true for any s_j such that $\mathbf{P}_{jq}(a) \neq 0$, then w is cyclically WMG-solvable.

Proof. We have proved in Lemma 2 that all SSPs are solvable for prime cycles. Let us consider the ESSPs at states s as in $w = \ldots \lfloor_s ab \ldots$, i.e. $\neg s[b\rangle$. Since we are looking for a WMG solution, all the sought places are of the form $p_{a,b}$ (see Lemma 3 and Fig. 4) with m, n as in (1). To define the initial marking of $p_{a,b}$, let us put $n \cdot \mathbf{P}(w)(b)$ tokens on it and fire the sequence w once completely. Choose some state s' in $w = \ldots \lfloor_{s'} a \ldots$ such that the number k of tokens on $p_{a,b}$ at state s' is minimal (it may be the case that such an s' is not unique; we can choose any such state). Define $M_0(p_{a,b}) = n \cdot \mathbf{P}(w)(b) - k$ as the initial marking of $p_{a,b}$. By construction, the firing of w reproduces the markings of

$p_{a,b}$ and M_0 guarantees their non-negativity. Let us show that the constructed place $p_{a,b}$ solves all the ESSPs $\neg s[b\rangle$, where $w = \dots \lfloor_s ab \dots$. Consider such a state s in w (w.l.o.g. we assume $s \neq s'$, since s' certainly disables b). From $w = u_1 \lfloor_{s'} a \dots \lfloor_s a b u_2$ (circularly) and from inequality (6) for $s_j = s'$ and $s_q = s$, we get $\mathbf{P}_{s's}(a) \cdot m - \mathbf{P}_{s's}(b) \cdot n < n$ since $\mathbf{P}_{s's}(a) > 0$. Since $M_{s'}(p_{a,b}) = 0$, $M_s(p_{a,b}) = M_{s'}(p_{a,b}) + \mathbf{P}_{s's}(a) \cdot m - \mathbf{P}_{s's}(b) \cdot n < n$, i.e., $p_{a,b}$ disables b at state s.

Now, we show that places of the form $p_{a,b}$ also solve the other ESSPs against b, i.e., at the states where b is not the subsequent transition. Sequence w (up to rotation) can be written as $w = u_1 b^{x_1} u_2 b^{x_2} \dots u_l b^{x_l}$, $1 \le l \le \mathbf{P}(w)(b)$, and for $1 \le i \le l$: $x_i > 0$, $u_i \in (T \setminus \{b\})^+$. Transition b has to be disabled at all the states between successive b-blocks. Consider an arbitrary pair of such blocks b^{x_j} and $b^{x_{j+1}}$ in $w = \dots b^{x_j} u_j b^{x_{j+1}} \dots = \dots b^{x_j} \lfloor_s u'_j \lfloor_r t b^{x_{j+1}} \dots$, with $u_j = u'_j t$. Place $p_{t,b}$ does not allow b to fire at state r. We have to check that b is not enabled at any state between s and r, i.e., it is not enabled 'inside' u'_j. If u'_j is empty, then $s = r$, and we are done. Let $u'_j \neq \varepsilon$. Due to $\mathbf{P}(u'_j)(b) = \mathbf{P}(u_j)(b) = 0$, the marking of place $p_{t,b}$ cannot decrease from s to r, i.e., $M_s(p_{t,b}) \le M_{s''}(p_{t,b}) \le M_r(p_{t,b})$ for any s'' 'inside' u'_j. Since $p_{t,b}$ disables b at r, it then disables b at all states between s and r, inclusively.   □

From Lemmas 4 and 5 we can deduce the following characterisation.

Theorem 4 (A characterisation of cyclic WMG-solvability). *A sequence* $w \in T^*$ *is cyclically solvable with a WMG iff* $\mathbf{P}(w)$ *is prime and, for any pair of circularly adjacent labels in* w, *for instance* $w = \dots \lfloor_{s_q} ab \dots$,

$$\frac{m}{n} < \frac{\mathbf{P}_{jq}(b) + 1}{\mathbf{P}_{jq}(a)}$$

holds true with m, n *as in (1) for any* s_j *such that* $\mathbf{P}_{jq}(a) \neq 0$. *A WMG solution can be found with the places of the form* $p_{a,b}$ *for every such pair of* a *and* b.

4.3   A Polynomial-Time Algorithm for Cyclic WMG-Solvability

From the characterisation given by Theorem 4 and the considerations above, Algorithm 1 below synthesises a cyclic WMG solution for a given sequence $w \in T^*$, if one exists.

The algorithm works as follows. Initially, the Parikh vector of the input sequence is calculated and checked for primeness in lines 2–3. If the Parikh vector is prime, we consecutively consider all the pairs of adjacent letters and examine the inequality from Theorem 4 for them. To achieve it, we take the first two letters in the sequence (lines 4–11), check if the inequality is satisfied for all the states (lines 12–18), and if so, construct a new place connecting the two letters under consideration (19–25). Then, the sequence is cyclically rotated such that the initial letter goes to the end and the second letter becomes initial (line 8). In the new sequence, we take again the first two letters (lines 9–11) and repeat the procedure for them. The algorithm stops after a complete rotation of the initial sequence, and by this moment all the pairs of adjacent letters have been checked. The ordered alphabet is stored in the array T, and the sequence is stored in v. We use variables a and b to store the letters under consideration in each step, ia and ib to store their indices in the alphabet, and na and nb are used for counting their occurrences during the check of the inequality from Theorem 4. Variables M and $Mmin$ are used to compute the initial marking of the sought place.

Algorithm 1: Synthesis of a WMG solving a cyclic word

    input : $w \in T^n$, $T = \{t_0, \dots, t_{m-1}\}$
    output: A WMG system (N, M_0) cyclically solving w, if it exists
1   var: $T[0 \mathbin{..} m - 1] = (t_0, \dots, t_{m-1})$, $v[0 \mathbin{..} n - 1]$, a, b, na, nb, ia, ib, M, $Mmin$;
2   compute the Parikh vector $\mathbf{P}[0 \mathbin{..} m - 1]$ of w;
3   **if** $\mathbf{P}$ *is not prime* **then return** *unsolvable* ;   // Parikh-primeness
4   $b \leftarrow w[0]$;
5   **for** $j = 0$ **to** $m - 1$ **do**   // index of b
6   **if** $b = T[j]$ **then** $ib \leftarrow j$

7   **for** $i = 0$ **to** $n - 1$ **do**
8      $v \leftarrow w[i] \dots w[n-1] w[0] \dots w[i-1]$;   // rotation of w
9      $a \leftarrow b$, $b \leftarrow v[1]$, $ia \leftarrow ib$;   // fix first adjacent pair
10   **for** $j = 0$ **to** $m - 1$ **do**
11   **if** $b = T[j]$ **then** $ib \leftarrow j$

12    $na \leftarrow 1$, $nb \leftarrow 1$;
13   **if** $a \neq b$ **then**
14   **for** $k = 2$ **to** $n - 1$ **do**
15   **if** $\frac{\mathbf{P}[ib]}{\mathbf{P}[ia]} \geq \frac{\mathbf{P}[ib] - nb + 1}{\mathbf{P}[ia] - na}$ **then**
16   **return** *unsolvable* ;   // check solvability condition
17   **if** $v[k] = T[ia]$ **then** $na \leftarrow na + 1$;
18   **if** $v[k] = T[ib]$ **then** $nb \leftarrow nb + 1$;

19       $M \leftarrow \mathbf{P}[ia] \cdot \mathbf{P}[ib]$, $Mmin \leftarrow M$;
20   **for** $k = 0$ **to** $n - 1$ **do**   // find initial marking
21   **if** $w[k] = a$ **then** $M \leftarrow M + \mathbf{P}[ib]$;
22   **if** $w[k] = b$ **then** $M \leftarrow M - \mathbf{P}[ia]$;
23   **if** $M < Mmin$ **then** $Mmin \leftarrow M$;

24   add new place $p_{T[ia], T[ib]}$ to N with
25       $W(T[ia], p) = \mathbf{P}[ib]$, $W(p, T[ib]) = \mathbf{P}[ia]$, $M_0 = \mathbf{P}[ia] \cdot \mathbf{P}[ib] - Mmin$;

26 **return** (N, M_0)

Polynomial-time Complexity of Algorithm 1. For a sequence of length n over an alphabet with m labels, the Parikh vector can be computed in $\mathcal{O}(n)$ and its primeness can be checked using e.g. the Euclidean algorithm, with a running time in $\mathcal{O}(m \cdot \log_2^2 n)$. The main for-cycle of Algorithm 1 involves the enumeration of all pairs of distinct states of the cycle. For each pair of adjacent labels, a run of the `for`-cycle consists of a lookup for an index in $\mathcal{O}(m)$, the verification of the inequality in $\mathcal{O}(n)$ and the construction of a place in $\mathcal{O}(n)$, which sums up to $\mathcal{O}(m+n)$. Thus, the main `for`-loop requires a runtime in $\mathcal{O}(n(n+m))$. Taking into account that $m \leq n$, and that n growths much faster than $\log_2^2 n$, the overall running time of the algorithm does not exceed $\mathcal{O}(n^2)$.

Complexity Comparison: the Known General Approach Is Less Efficient. For a comparison, solving a cycle of length n over m labels with a WMG amounts to solve $n(n-1)$ SSPs and $n(m-1)$ ESSPs at most. In the special case of WMG synthesis from a prime cycle, we know that all the SSPs are solvable (Lemma 2) and that solving first the other problems avoids to consider the SSPs (see [21]). Since each of the sought places has at most one input and one output, each of the separation problems seeks for 3 unknown variables, namely the initial marking of a place, the input and the output arc weights. For an ESSP, the output transition is clearly the one which has to be disabled and the input transition is to be found. So, there are $m-1$ possibilities to define a concrete ESSP, which in the worst case gives us up to $n(m-1)^2$ systems of inequalities to solve all the ESSPs.

The general region-based synthesis typically uses ILP-solvers, and using e.g. Karmarkar's algorithm [31] (which is known to be efficient) for solving an ILP-problem with k unknowns, we expect a running time of $\mathcal{O}(k^{3.5} \cdot L^2 \cdot \log L \cdot \log \log L)$ where L is the length of the input in bits. For the case of a cycle, the input of each separation problem is the matrix with the range of $(n+1) \times m$ and the vector of right sides with the range of $n+1$, where each component of the vector and of the matrix is a natural number not greater than n. Hence, the length of the input for a single separation problem can be estimated as $L = (m+1) \cdot (n+1) \cdot \log_2 n$ bits, implying a runtime of $\mathcal{O}(n^2 \cdot m^2 \cdot \log_2^2 n \cdot \log L \cdot \log \log L)$ for solving a single separation problem (the number of unknowns being equal to 3, i.e. constant). Thus the general synthesis approach would need a runtime of $\mathcal{O}(n^3 \cdot m^4 \cdot LF)$ with the logarithmic factor $LF = \log_2^2 n \cdot \log((m+1) \cdot (n+1) \cdot \log_2 n) \cdot \log \log((m+1) \cdot (n+1) \cdot \log_2 n)$. Note that, with this general approach, some redundant places may be constructed, but they can be wiped out in a post-processing phase.

4.4   CF-Solvability vs WMG-Solvability of Cycles

Let us now relate cyclic WMG-solvability to cyclic CF-solvability.

Theorem 5. *A sequence $u \in \{a, b, c\}^*$ is cyclically WMG-solvable iff u is cyclically CF-solvable.*

Proof. WMGs form a proper subclass of CF nets, hence the direct implication. Let now $TS = (S, T = \{a, b, c\}, \rightarrow, s_0)$ be a CF-solvable circular LTS obtained from u and $\Upsilon = \mathbf{P}(u)$. By contraposition, assume that TS is not solvable by a WMG. Then, due to Theorem 4, for some distinct states $j, q \in S$ and distinct labels $a, b \in T$

$$\frac{\Upsilon(a)}{\Upsilon(b)} \geq \frac{\mathbf{P}_{jq}(a) + 1}{\mathbf{P}_{jq}(b)}. \tag{7}$$

W.l.o.g. we can choose the leftmost j satisfying (7). Then, in TS we have $r[a\rangle j$ for some $r \in S$ preceding j. Indeed, if this is not the case and either $r[b\rangle j$ or $r[c\rangle j$, then (7) holds true for r and q, contradicting the choice of j. On the other hand, since $\mathbf{P}_{jq}(a) + 1 = \mathbf{P}_{rq}(a)$ and $\mathbf{P}_{jq}(b) = \mathbf{P}_{rq}(b)$, the inequality (7) implies

$$\frac{\Upsilon(a)}{\Upsilon(b)} \geq \frac{\mathbf{P}_{rq}(a)}{\mathbf{P}_{rq}(b)}. \tag{8}$$

Consider a place p which is an input place of a in a cyclic CF solution of u. From Lemma 1, we can assume pureness, i.e., the place has the form illustrated on the right of Fig. 5 with $x = a$, $y = b$, $z = c$.

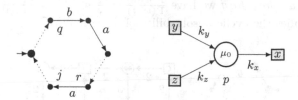

Fig. 5. u (left) is cyclically solvable by a CF system; a CF place over $\{x, y, z\}$ (right).

We must have the following constraints for p:

$$
\begin{aligned}
\text{cycle} \quad &: k_b \cdot \Upsilon(b) + k_c \cdot \Upsilon(c) = k_a \cdot \Upsilon(a) \\
r[a\rangle \quad &: M_r(p) \geq k_a \\
r[a \ldots\rangle q \quad &: M_q(p) = M_r(p) + k_b \cdot \mathbf{P}_{rq}(b) + k_c \cdot \mathbf{P}_{rq}(c) - k_a \cdot \mathbf{P}_{rq}(a).
\end{aligned} \tag{9}
$$

If $\mathbf{P}_{rq}(c) \geq \mathbf{P}_{rq}(a) \cdot \frac{\Upsilon(c)}{\Upsilon(a)}$, then due to (8) and (9),

$$
\begin{aligned}
M_q(p) = \quad & M_r(p) + k_b \cdot \mathbf{P}_{rq}(b) + k_c \cdot \mathbf{P}_{rq}(c) - k_a \cdot \mathbf{P}_{rq}(a) \\
\geq \quad & k_a + \left(k_b \cdot \frac{\Upsilon(b)}{\Upsilon(a)} + k_c \cdot \frac{\Upsilon(c)}{\Upsilon(a)} - k_a\right) \cdot \mathbf{P}_{rq}(a) = k_a,
\end{aligned}
$$

implying $q[a\rangle$ which contradicts $q[b\rangle$. Hence, $\mathbf{P}_{rq}(c) < \mathbf{P}_{rq}(a) \cdot \frac{\Upsilon(c)}{\Upsilon(a)}$. Together with (8), we have

$$\frac{\mathbf{P}_{rq}(b)}{\Upsilon(b)} \geq \frac{\mathbf{P}_{rq}(a)}{\Upsilon(a)} > \frac{\mathbf{P}_{rq}(c)}{\Upsilon(c)}.$$

which is equivalent to

$$\frac{\mathbf{P}_{qr}(b)}{\Upsilon(b)} \leq \frac{\mathbf{P}_{qr}(a)}{\Upsilon(a)} < \frac{\mathbf{P}_{qr}(c)}{\Upsilon(c)}. \tag{10}$$

For an arbitrary input place of b, hence of the form illustrated on the right of Fig. 5 with $x = b$, $y = a$, $z = c$,

$$
\begin{aligned}
\text{cycle} \quad &: k_a \cdot \Upsilon(a) + k_c \cdot \Upsilon(c) = k_b \cdot \Upsilon(b) \\
q[b\rangle \quad &: M_q(p) \geq k_b \\
q[b \ldots\rangle r \quad &: M_r(p) = M_q(p) + k_a \cdot \mathbf{P}_{qr}(a) + k_c \cdot \mathbf{P}_{qr}(c) - k_b \cdot \mathbf{P}_{qr}(b).
\end{aligned} \tag{11}
$$

Then, due to (10) and (11),

$$M_r(p) = M_q(p) + k_a \cdot \mathbf{P}_{qr}(a) + k_c \cdot \mathbf{P}_{qr}(c) - k_b \cdot \mathbf{P}_{qr}(b)$$
$$\geq k_b + \left(k_a \cdot \frac{\Upsilon(a)}{\Upsilon(b)} + k_c \cdot \frac{\Upsilon(c)}{\Upsilon(b)} - k_b \right) \cdot \mathbf{P}_{qr}(b) = k_b$$

which implies $r[b\rangle$, a contradiction. Thus, TS is solvable by some WMG.   □

When the alphabet has more than three elements, the inclusion of WMGs into CF nets is strict, i.e., there are cyclically CF-solvable sequences that are not cyclically WMG-solvable: the sequence $w = abcbad$ has a cyclic CF solution (cf. Fig. 6); for $a \lfloor_r bc \lfloor_q bad$ we have $\frac{\mathbf{P}(w)(a)}{\mathbf{P}(w)(b)} = \frac{2}{2} \not< \frac{0+1}{1} = \frac{\mathbf{P}_{rq}(a)+1}{\mathbf{P}_{rq}(b)}$ which, by Theorem 4, implies the cyclic unsolvability of w by a WMG.

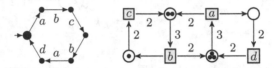

Fig. 6. Sequence $abcbad$ is cyclically solved by the CF system on the right.

By Lemma 3, using places only between adjacent transitions is sufficient for cyclic WMG-solvability. For the sequence $abcbad$ in Fig. 6, b follows a and c, and the input place of b in the CF solution is an output place for both a and c. The situation is similar for a, which follows b and d. However, this is not always the case when we are looking for a solution in the class of CF nets. For instance, the sequence $cabdaaab$ is cyclically solvable by a CF system (see Fig. 7). In this sequence, b always follows a. But in order to solve ESSPs against b, we need an output place for c (in addition to a). Indeed, if there is a place $p_{a,b}$ as on the right of Fig. 7 which solves ESSPs against b, then for $ca \lfloor_s bdaa \lfloor_q ab$ we get

$$s[b\rangle \quad : \mu_0 + 2 \qquad\qquad \geq 4$$
$$\neg q[b\rangle \quad : \mu_0 + 3 \cdot 2 - 4 \quad < 4$$

Subtracting the first inequality from the second one, we get $4 - 4 < 0$, a contradiction. Hence, $p_{a,b}$ cannot solve all ESSPs against b in the cycle $cabdaaab$.

In a WMG, a place has at most one input. This restriction is relaxed for CF nets: multiple inputs are allowed. Let us show that a single input place for each transition is not always sufficient. For instance, consider the cyclically CF-solvable sequence $bcafdeaaabcdaafdcaaa$ and Fig. 8. Assume we can solve all ESSPs against transition a with a single place p as on the right of the same figure; due to Lemma 1, we do not need any side-condition. Then, for p and $w = \lfloor_{s_0} bcafd \lfloor_{s_5} e \lfloor_{s_6} aaabc \lfloor_{s_{11}} d \lfloor_{s_{12}} aafd \lfloor_{s_{16}} c \lfloor_{s_{17}} aaa$, the following system of inequalities must hold true:

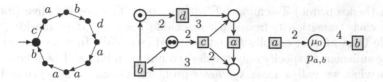

Fig. 7. *cabdaaab* is cyclically CF-solvable (middle), but is not cyclically WMG-solvable.

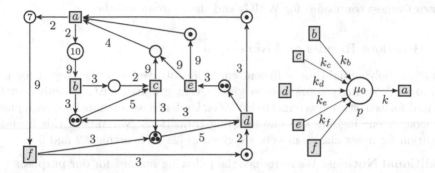

Fig. 8. $w = bcafdeaaabcdaafdcaaa$ is cyclically solved by the CF system on the left; a (pure) place of a CF system with 6 transitions on the right.

$$
\begin{array}{llll}
\text{cycle} & : 2 \cdot k_b + 3 \cdot k_c + 3 \cdot k_d + k_e + 2 \cdot k_f & = 9 \cdot k & (0) \\
\neg s_5[a\rangle & : \mu_0 + k_b + k_c + k_d + k_f - k & < k & (1) \\
s_6[aaa\rangle & : \mu_0 + k_b + k_c + k_d + k_e + k_f - k & \geq 3 \cdot k & (2) \\
\neg s_{11}[a\rangle & : \mu_0 + 2 \cdot k_b + 2 \cdot k_c + k_d + k_e + k_f - 4 \cdot k & < k & (3) \\
s_{12}[aa\rangle & : \mu_0 + 2 \cdot k_b + 2 \cdot k_c + 2 \cdot k_d + k_e + k_f - 4 \cdot k & \geq 2 \cdot k & (4) \\
\neg s_{16}[a\rangle & : \mu_0 + 2 \cdot k_b + 2 \cdot k_c + 3 \cdot k_d + k_e + 2 \cdot k_f - 6 \cdot k & < k & (5) \\
s_{17}[aaa\rangle & : \mu_0 + 2 \cdot k_b + 3 \cdot k_c + 3 \cdot k_d + k_e + 2 \cdot k_f - 6 \cdot k & \geq 3 \cdot k & (6)
\end{array}
$$

From the system above we obtain:

$$
\begin{array}{lll}
(2) - (1) & : k_e & > 2 \cdot k \\
(4) - (3) & : k_d & > k \\
(6) - (5) & : k_c & > 2 \cdot k
\end{array}
$$

which implies $3 \cdot k_c + 3 \cdot k_d + k_e > 11 \cdot k$, contradicting the equality (0). Hence, the ESSPs against a cannot be solved by a single place.

5 Weak Synthesis of WMGs in Polynomial-Time

For any given word w over a set of labels T whose support equals T, each system $\mathcal{S} = ((P, T, W), M_0)$ that cyclically WMG-solves w, when it exists, has a unique minimal (hence prime) T-semiflow Υ with support T, since it is live (meaning that for each transition t, from each reachable marking M, a marking M' is reachable from M that enables t) and bounded (see [35]). In some situations, it might be sufficient to specify only the desired unique minimal T-semiflow, which leads to what we call a *weak synthesis* problem. Given such a prime Parikh vector Υ, the aim is thus to construct a WMG cyclically solving an arbitrary sequence whose Parikh vector equals Υ. In this section, we provide a method for constructing a solution in polynomial-time. To achieve it, we first need to recall known liveness conditions for WMGs and their circuit subclass.

5.1   Previous Results on Liveness

In [32], a polynomial-time sufficient condition of liveness is developed for the well-formed, strongly connected weighted event graphs (WEGs), equivalent to the well-formed, strongly connected WMGs. Under these assumptions, each place has exactly one ingoing and one outgoing transitions. Variants of this liveness condition for other classes of nets are given in [28], Theorems 4.2 and 5.5.

Additional Notions. We introduce the following notions for our purpose:

- For any place p, $\gcd_p$ denotes the gcd of all input and output weights of p.
- A marking M_0 satisfies the *useful tokens condition* if, for each place p, $M_0(p)$ is a multiple of $\gcd_p$. Indeed, if $M_0(p) = k \cdot \gcd_p + r$ for some non-negative integers k and r such that $0 < r < \gcd_p$, then r tokens are never used by any firing (see [28,32] for more details).
- A net (P, T, W) with incidence matrix I is *conservative* if there is a P-vector $X \geq \mathbb{1}^{|P|}$ such that $X \cdot I = \mathbb{0}^{|T|}$, where $\mathbb{1}^{|P|}$ denotes the vector of size $|P|$ in which each component has value 1. Such a P-vector X is called a *conservativeness* vector. The net is 1-*conservative* if $\mathbb{1}^{|P|}$ is a conservativeness vector, i.e. if for each transition, the sum of its input weights equals the sum of its output weights.
- A net N is *structurally bounded* if for each marking M_0, (N, M_0) is bounded.
- By Theorem 4.11 in [35], a live WTS (N, M_0) is bounded iff N is conservative. We focus on live and bounded WMG solutions (which are WTS), hence on conservative, thus structurally bounded (see [34]), solutions.
- The scaling operation (Definition 3.1 in [28]): The multiplication of all input and output weights of a place p together with its marking by a positive rational number α_p is a *scaling* of the place p if the resulting input and output weights and marking are integers. If each place p of a system is scaled by a positive rational α_p, the system is said to be scaled by the vector α whose components are the scaling factors α_p.

Recalling Theorem 3.2 in [28], if $\mathcal{S} = ((P, T, W), M_0)$ is a system and α is a vector of $|P|$ positive rational components, then scaling $\mathcal{S}$ by α preserves the feasible sequences of firings. We deduce from it, and from Theorem 3.5 in the same paper, the following result.

Lemma 6. *Consider a system S, whose scaling by a vector α of positive rationals yields the system S'. Then, for each feasible sequence σ, denote by M the marking reached when firing σ in S and by M' the marking reached when firing σ in S'; the set of places enabled by M equals the set of places enabled by M'.*

Thus, in conservative systems, we can reason equivalently on feasible sequences and enabled places in the system scaled by a conservativeness vector, yielding a 1-conservative system whose tokens amount remains constant.

The next result is a specialisation of Theorem 4.5 in [28] to circuit Petri nets, using the fact that the liveness of circuits is monotonic, i.e. preserved upon any addition of initial tokens (see [14, 28, 35] and Theorem 7.10 in [30] with its proof).

Proposition 1 (Sufficient condition of liveness [28, 30, 32]). *Consider a conservative circuit system $S = (N, M_0)$, with $N = (P, T, W)$. S is live if the following conditions hold:*

- *for a place p_0, with $\{t_0\} = p_0^\bullet$, $M_0(p_0) = W(p_0, t_0)$;*
- *for every place p in $P \setminus \{p_0\}$, with $p^\bullet = \{t\}$, $M_0(p) = W(p, t) - \gcd_p$.*

Moreover, for every marking M_0' such that $M_0' \geq M_0$, (N, M_0') is live.

In the particular case of a binary circuit, i.e. with two transitions, we recall the next characterisation condition of liveness, given as Theorem 5.2 in [32].

Proposition 2 (Liveness of binary 1-conservative circuits [32]). *Consider a 1-conservative binary circuit $S = ((P, T, W), M_0)$ that fulfills the useful tokens condition, with $T = \{a, b\}$ and $P = \{p_{a,b}, p_{b,a}\}$, where $p_{a,b}$ is the output of a and $p_{b,a}$ is the output of b. Let $m = W(a, p_{a,b})$ and $n = W(p_{a,b}, b)$. Then S is live iff $M_0(p_{a,b}) + M_0(p_{b,a}) > m + n - 2 \cdot \gcd(m, n)$.*

Now, consider any 1-conservative binary circuit system S whose initial marking M_0 marks one place $p_{a,b}$ with its output weight $W(p_{a,b}, b)$ and the other place $p_{b,a}$ with $W(p_{b,a}, a) - \gcd_{p_{b,a}}$. Each marking reachable from M_0 enables exactly one place; applying Lemma 6, this is also the case for any scaling of S, hence:

Lemma 7 (One enabled place in binary circuits). *Consider a conservative binary circuit system $S = (N, M_0)$, with $N = (P, T, W)$, $T = \{a, b\}$, such that for the place $p_{a,b}$ with output b, $M_0(p_{a,b}) = W(p_{a,b}, b)$ and for the other place $p_{b,a}$ with output a, $M_0(p_{b,a}) = W(p_{b,a}, a) - \gcd_{p_{b,a}}$. Then each reachable marking enables exactly one place.*

This lemma will help ensuring that the reachability graph of the synthesised WMG forms a circle. Figure 9 illustrates Lemma 7 and Proposition 1.

We recall a liveness characterisation. Since a place with an output and no input, called a *source-place*, prevents liveness, we assume there is no such place.

Proposition 3 (Liveness of WMGs [35]). *Consider a WMG S without source places. Then S is live iff each circuit P-subsystem of S is live.*

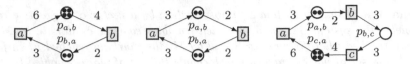

Fig. 9. These circuits are conservative and fulfill the sufficient condition of liveness of Proposition 1. On the left, $\gcd_{p_{a,b}} = 2$ and $\gcd_{p_{b,a}} = 1$. The system in the middle is obtained from the one on the left by scaling $p_{a,b}$ with $\frac{1}{2}$. In this second system, $\gcd_{p_{a,b}} = 1$ and $\gcd_{p_{b,a}} = 1$. On the right, $\gcd_{p_{a,b}} = 1$, $\gcd_{p_{b,c}} = 3$ and $\gcd_{p_{c,a}} = 2$.

5.2   Weak Synthesis of WMGs in Polynomial-Time

Algorithm 2 below constructs a WMG from a given prime T-vector Υ. We prove it terminates and computes a WMG cyclically solving some word with Parikh vector Υ, hence performing weak synthesis. We then show it lies in PTIME.

Algorithm 2: Weak synthesis of a WMG with circular RG.

Data: A prime T-vector Υ with support $T = \{t_1, \ldots, t_m\}$.
Result: A WMG cyclically solving a word with Parikh vector Υ.

1 We construct first an unmarked WMG $N = (P, T, W)$ containing all possible binary circuits (which we call the complete WMG), as follows:

2 **for** *each pair of distinct labels t_i, t_j in T* **do**

3   Add two new places $p_{i,j}$ and $p_{j,i}$ forming a binary circuit P-subnet with set of labels $\{t_i, t_j\}$, such that:

4     $W(p_{i,j}, t_j) = W(t_j, p_{j,i}) = \frac{\Upsilon(t_i)}{\gcd(\Upsilon(t_i), \Upsilon(t_j))}$

5     $W(p_{j,i}, t_i) = W(t_i, p_{i,j}) = \frac{\Upsilon(t_j)}{\gcd(\Upsilon(t_i), \Upsilon(t_j))}$.

6 Then, we construct its initial marking M_0, visiting the transitions in increasing order, as follows:

7 **for** $i = 2..m$ **do**

8   Mark each output place $p_{i,h}$ of t_i that is an input of a transition t_h of smaller index, i.e. $h < i$, with $M_0(p) = W(p_{i,h}, t_h)$;

9   Mark each input place $p_{h,i}$ of t_i that is an output of a transition t_h of smaller index, i.e. $h < i$, with
   $M_0(p) = W(p_{h,i}, t_i) - \gcd_{p_{h,i}} = W(p_{h,i}, t_i) - 1$;

10 **return** (N, M_0)

Theorem 6 (Weak synthesis of a WMG). *For every prime T-vector Υ, Algorithm 2 terminates and computes a WMG cyclically solving Υ, i.e. cyclically solving some word $w \in T^*$ such that $\mathbf{P}(w) = \Upsilon$.*

Proof. The proof is illustrated in Figs. 10, 11 and 12. Consider any prime T-vector $\Upsilon \in (\mathbb{N} \setminus \{0\})^m$, where m is the number of transitions. In the first loop, we consider each pair of transitions once. In the second loop, we consider each place once. Thus, the algorithm terminates. Let us prove its correction.

If $|T| = 1$, there is one transition and $\Upsilon = (1)$: the WMG with $T = \{t_1\}$, $P = \emptyset$ and the sequence $w = t_1$ fulfill the claim. Hence, we suppose $|T| \geq 2$.

For each place $p_{i,j}$, we have $\Upsilon(t_j) \cdot W(p_{i,j}, t_j) = \Upsilon(t_i) \cdot W(t_i, p_{i,j})$, so that $-W(p_{i,j}, t_j) \cdot \Upsilon(t_j) + W(t_i, p_{i,j}) \cdot \Upsilon(t_i) = 0$, hence $I \cdot \Upsilon = 0$, where I is the incidence matrix of N. Moreover, each circuit P-subnet of N is conservative (by Corollary 3.6 in [35]).

Now, let us consider the second loop: we prove the next invariant $\mathrm{Inv}(\ell)$ to be true at the end of each iteration ℓ, for each $\ell = 1..m - 1$, by induction on ℓ:

$\mathrm{Inv}(\ell)$: "At the end of the ℓ-th iteration, the WMG P-subsystem S_ℓ defined by the set of places $P_\ell = \{p_{u,v} \mid u, v \in \{1, \ldots, \ell+1\}, u \neq v\}$ is live, and each binary circuit P-subsystem of S_ℓ has exactly one enabled place".

Before entering the loop, i.e. before the first iteration, the WMG in unmarked.

Base case: $\ell = 1$. At the end of the first iteration, $P_\ell = P_1 = \{p_{1,2}, p_{2,1}\}$, which induces a live binary circuit (by Proposition 1) with exactly one enabled place, since only one output of t_2 is enabled by M_0 and the other place is an output of t_1 considered in the second part of the loop.

Inductive case: $1 < \ell \leq m - 1$. We suppose $\mathrm{Inv}(\ell-1)$ to be true, and we prove that $\mathrm{Inv}(\ell)$ is true. Thus, at the end of iteration $\ell - 1$, we suppose that the P-subsystem $S_{\ell-1}$ induced by $P_{\ell-1}$ is live, and that each binary circuit P-subsystem of $S_{\ell-1}$ has exactly one enabled place. The iteration ℓ marks only all the input and output places of $t_{\ell+1}$ that are inputs or outputs of transitions in $\{t_1, \ldots, t_\ell\}$. None of these places has been considered in any previous iteration, since each iteration considers only places connected to transitions of smaller index. Thus, these places are newly marked at iteration ℓ, and the only places unmarked at the end of this iteration are connected to transitions of higher index.

We deduce that, at the end of iteration ℓ:

– each binary circuit of S_ℓ has exactly one enabled place: indeed, each such binary circuit either belongs to $S_{\ell-1}$, on which the inductive hypothesis applies, or to the circuits newly marked at iteration ℓ;

– each circuit P-subsystem of S_ℓ with three places or more is live: indeed, consider any such conservative circuit C; either C is a P-subsystem of $S_{\ell-1}$, which is live by the inductive hypothesis, hence Proposition 3 applies and C is live, or C contains transition $t_{\ell+1}$, in which case C contains necessarily an output $p_{\ell+1,h}$ of $t_{\ell+1}$ with $h < \ell + 1$: since each place $p_{u,v}$ of S_ℓ is marked with at least $W(p_{u,v}, t_v) - \gcd_{p_{u,v}}$ and $p_{\ell+1,h}$ is marked with $W(p_{\ell+1,h}, t_h)$, C fulfills the sufficient condition of liveness of Proposition 1, hence is live; we deduce that each circuit P-subsystem is live, hence S_ℓ is live by Proposition 3.

We proved that $\mathrm{Inv}(\ell)$ is true for every integer $\ell = 1..m - 1$. We deduce that the WMG system $S_{m-1} = (N, M_0)$ obtained at the end of the last iteration, which is the system returned, fulfills $\mathrm{Inv}(m - 1)$. Suppose that some marking M reachable in S_{n-1} enables two distinct transitions t_i and t_j. Since S_{n-1} is a complete WMG, there is a binary circuit P-subsystem $C_{i,j} = (N_{i,j}, M_{i,j})$ containing t_i and t_j, in which exactly one place is enabled, applying Lemma 7 (since $M|_{\{p_{i,j},p_{j,i}\}} = M_{i,j}$ is a marking reachable in $(N_{i,j}, M_0|_{\{p_{i,j},p_{j,i}\}})$). We deduce that M cannot enable both t_i and t_j, a contradiction.

Thus the WMG returned is live and each of its reachable markings enables exactly one transition. It is known that, in each live and bounded WMG, a sequence σ is feasible such that $\mathbf{P}(\sigma) = \Upsilon$ which is the unique minimal T-semiflow of the WMG (see [35,36]). Consequently, its reachability graph is a circle, i.e. the WMG solves Υ (and σ) cyclically. We get the claim.   □

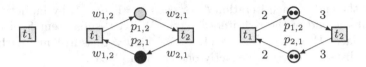

Fig. 10. Sketching Theorem 6 for 1 and 2 transitions. On the left, the circuit system $\mathcal{S}_1$ has no place and solves $\Upsilon = (1)$. In the circuit system $\mathcal{S}_2$ in the middle, the output of t_2 is marked as black and its input as grey. On the right, an instanciation of the binary case, given $\Upsilon = (3, 2)$. These systems are live, $RG(\mathcal{S}_1)$ and $RG(\mathcal{S}_2)$ are circles.

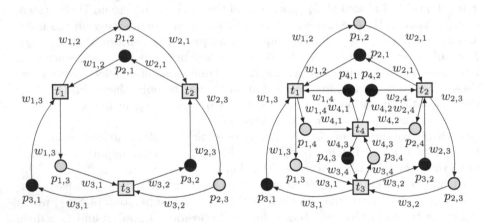

Fig. 11. Sketching Theorem 6 for 3 and 4 transitions (systems $\mathcal{S}_3, \mathcal{S}_4$). Each black place $p_{i,j}$ is marked with $W(p_{i,j}, t_j)$, each grey place $p_{i,j}$ is marked with $W(p_{i,j}, t_j) - \gcd_{p_{i,j}}$. On the left, in the circuit induced by $\{p_{1,2}, p_{2,1}\}$, the output of t_2 is black and its input is grey. Then, each output of t_3 is black, each input is grey. Each circuit of $\mathcal{S}_3$ is live, $\mathcal{S}_3$ is live and $RG(\mathcal{S}_3)$ is a circle. In $\mathcal{S}_4$, we keep the marking of $\mathcal{S}_3$ and mark each output of t_4 as black, each of its inputs as grey. Thus, $\mathcal{S}_4$ is live and $RG(\mathcal{S}_4)$ is a circle.

Polynomial-time Complexity of Algorithm 2. Let m be the number of transitions (labels). The initial construction of the net N considers a number of transition pairs equal to $m \cdot (m-1)$. The computation of $\gcd(i, j)$ for any two integers i, j can be done using the Euclidean algorithm in $\mathcal{O}(log_2^2(\max(i, j)))$, which

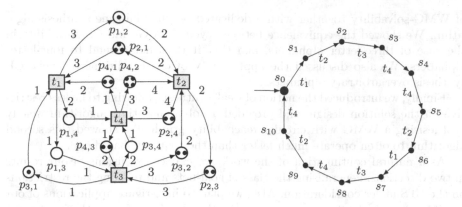

Fig. 12. Illustration of the proof of Theorem 6 for the prime T-vector $\varUpsilon = (2,3,2,4)$. On the left, a complete WMG $\mathcal{S}$, with all possible binary circuits. Its marking follows the black and grey places of Fig. 11. Pick any circuit P-subsystem C, e.g. the one induced by $\{p_{4,3}, p_{3,2}, p_{2,4}\}$: it is conservative and fulfills the condition of Proposition 1, hence it is live. On the right, an LTS representing $RG(\mathcal{S})$. The sequence $w = t_1\, t_2\, t_3\, t_4\, t_4\, t_2\, t_1\, t_3\, t_4\, t_2\, t_4$, with $\mathbf{P}(w) = \varUpsilon$, is cyclically WMG-solvable.

remains polynomial in the size of the input vector binary encoding. Computing $\frac{u}{v}$ for two integers u, v, $v \neq 0$, can also be done in $\mathcal{O}(log_2^2(\max(u,v)))$. Thus, constructing N lies in $\mathcal{O}(m(m-1)3log_2^2(q))$ where q is the highest value in $\varUpsilon$. Then, the algorithm marks all places in $\mathcal{O}(m(m-1))$, knowing that the gcd of each place is 1. Hence the algorithm lies in PTIME: $\mathcal{O}(3m(m-1)log_2^2(q)+m(m-1))$, i.e. $\mathcal{O}(m(m-1)(3log_2^2(q)+1))$ where q is the highest value in $\varUpsilon$.

Comparison with the Sequence-Based Synthesis. Algorithm 1 uses $\mathcal{O}(n(m+n))$ steps, where m is the number of labels and n is the length of the input sequence w to be solved cyclically. Since n equals the sum of the components of $\varUpsilon = \mathbf{P}(w)$, we get $q \leq n$. Also, $m \leq n$; depending on the weights, n can be exponentially larger than m. Hence, $log_2^2(q) \in \mathcal{O}(log_2^2(n))$, so that $m(m-1)(3log_2^2(q)+1) \in \mathcal{O}(m^2 \cdot log_2^2(n))$. When n is exponential in m, Algorithm 2 operates in time polynomial in m while Algorithm 1 operates in time exponential in m.

6   Conclusions and Perspectives

In this work, we specialised previous methods of analysis and synthesis to the CF nets and their WMG subclass, two useful subclasses of weighted Petri nets allowing to model various real-world applications.

We highlighted the correspondence between CF- and WMG-solvability for binary alphabets. We also tackled the case of an LTS formed of a single cycle with an arbitrary number of letters, for which we developed a characterisation of WMG-solvability together with a dedicated polynomial-time synthesis algorithm. We showed the equivalence between cyclic WMG- and CF-solvability in the case of three-letter alphabets, and that it does not extend to four-letter alphabets. We also discussed the applicability of our conditions to cyclic CF synthesis over arbitrary alphabets.

Finally, we introduced the notion of weak synthesis, allowing to be less restrictive on the solution design, and provided a polynomial-time algorithm weakly synthesising a WMG with circular reachability graph. We showed this second algorithm to often operate much faster than the sequence-based one.

As a natural continuation of the work, we expect extensions of our results in two directions: generalising the class of goal-nets and relaxing the restrictions on the LTS under consideration. Also, we plan to investigate applications of our results, in particular to *process mining*[1]: the algorithm for synthesising a net from a transition system can be applied to classical *process discovery* which is also a kind of synthesis; and the algorithm for weak synthesis can be employed in *streaming process mining*, where only partial information about the behaviour of the modelled system can be stored for a later processing.

Acknowledgements. We would like to thank the anonymous referees for their careful proofreading, their relevant comments and their insightful suggestion of addressing the process mining area.

References

1. van der Aalst, W.M.P.: Process Mining - Data Science in Action. Second Edition. Springer (2016). https://doi.org/10.1007/978-3-662-49851-4
2. Petri Net Synthesis. TTCSAES. Springer, Heidelberg (2015). https://doi.org/10.1007/978-3-662-47967-4_14
3. Badouel, E., Bernardinello, L., Darondeau, P.: Polynomial algorithms for the synthesis of bounded nets. In: Mosses, P.D., Nielsen, M., Schwartzbach, M.I. (eds.) CAAP 1995. LNCS, vol. 915, pp. 364–378. Springer, Heidelberg (1995). https://doi.org/10.1007/3-540-59293-8_207
4. Badouel, E., Bernardinello, L., Darondeau, P.: The synthesis problem for elementary net systems is NP-complete. Theoret. Comput. Sci. **186**(1), 107–134 (1997). https://doi.org/10.1016/S0304-3975(96)00219-8
5. Barylska, K., Best, E., Erofeev, E., Mikulski, L., Piatkowski, M.: On binary words being Petri net solvable. In: Proceedings of the International Workshop on Algorithms & Theories for the Analysis of Event Data, Brussels, Belgium, pp. 1–15 (2015)
6. Barylska, K., Best, E., Erofeev, E., Mikulski, L., Piatkowski, M.: Conditions for Petri net solvable binary words. T. Petri Nets Other Models Concurrency **11**, 137–159 (2016). https://doi.org/10.1007/978-3-662-53401-4_7
7. Best, E., Devillers, R.: Synthesis and reengineering of persistent systems. Acta Informatica **52**(1), 35–60 (2014). https://doi.org/10.1007/s00236-014-0209-7
8. Best, E., Devillers, R.: Characterisation of the state spaces of marked graph Petri nets. Inf. Comput. **253**(3), 399–410 (2017)
9. Best, E., Devillers, R., Schlachter, U.: Bounded choice-free Petri net synthesis: Algorithmic issues. Acta Informatica (2017)

10. Best, E., Devillers, R., Schlachter, U., Wimmel, H.: Simultaneous Petri net synthesis. Sci. Ann. Comp. Sci. **28**(2), 199–236 (2018)
11. Best, E., Hujsa, T., Wimmel, H.: Sufficient conditions for the marked graph realisability of labelled transition systems. Theoretical Comput. Sci. **750**, 101–116 (2017)
12. Commoner, F., Holt, A., Even, S., Pnueli, A.: Marked directed graphs. J. Comput. Syst. Sci. **5**(5), 511–523 (1971). https://doi.org/10.1016/S0022-0000(71)80013-2
13. Crespi-Reghizzi, S., Mandrioli, D.: A decidability theorem for a class of vector-addition systems. Inf. Process. Lett. **3**(3), 78–80 (1975). https://doi.org/10.1016/0020-0190(75)90020-4
14. Delosme, J.M., Hujsa, T., Munier-Kordon, A.: Polynomial sufficient conditions of well-behavedness for weighted join-free and choice-free systems. In: 13th International Conference on Application of Concurrency to System Design, pp. 90–99, July 2013. https://doi.org/10.1109/ACSD.2013.12
15. Desel, J., Esparza, J.: Free Choice Petri Nets, Cambridge Tracts in Theoretical Computer Science, vol. 40. Cambridge University Press, New York (1995)
16. Devillers, R.: Products of transition systems and additions of Petri Nets. In: Desel, J., Yakovlev, A. (eds) Proceedings of the 16th International Conference on Application of Concurrency to System Design (ACSD 2016), pp. 65–73 (2016)
17. Devillers, R.: Factorisation of transition systems. Acta Informatica **55**(4), 339–362 (2017). https://doi.org/10.1007/s00236-017-0300-y
18. Devillers, R.: Articulation of transition systems and its application to Petri net synthesis. In: Application and Theory of Petri Nets and Concurrency - 40th International Conference, Aachen, Germany, 23–28 June, 2019, Proceedings, pp. 113–126 (2019). https://doi.org/10.1007/978-3-030-21571-2_8
19. Devillers, R., Erofeev, E., Hujsa, T.: Synthesis of weighted marked graphs from constrained labelled transition systems. In: Proceedings of the International Workshop on Algorithms & Theories for the Analysis of Event Data, Satellite event of the Conferences: Petri Nets and ACSD, Bratislava, Slovakia, pp. 75–90 (2018)
20. Devillers, R., Erofeev, E., Hujsa, T.: Synthesis of weighted marked graphs from circular labelled transition systems. In: Proceedings of the International Workshop on Algorithms & Theories for the Analysis of Event Data, Satellite event of the conferences: Petri Nets and ACSD, Aachen, Germany, pp. 6–22, June 2019
21. Devillers, R., Hujsa, T.: Analysis and synthesis of weighted marked graph petri nets. In: Khomenko, V., Roux, O.H. (eds.) PETRI NETS 2018. LNCS, vol. 10877, pp. 19–39. Springer, Cham (2018). https://doi.org/10.1007/978-3-319-91268-4_2
22. Devillers, R., Hujsa, T.: Analysis and synthesis of weighted marked graph Petri nets: Exact and approximate methods. Fundamenta Informaticae (2019)
23. Devillers, R., Schlachter, U.: Factorisation of Petri net solvable transition systems. In: Application and Theory of Petri Nets and Concurrency - 39th International Conference, Bratislava, Slovakia, 24–29 June, 2018, Proceedings, pp. 82–98 (2018). https://doi.org/10.1007/978-3-319-91268-4_5
24. Erofeev, E., Barylska, K., Mikulski, L., Piatkowski, M.: Generating all minimal Petri net unsolvable binary words. In: Proceedings of the Prague Stringology Conference 2016, Prague, Czech Republic, pp. 33–46 (2016)
25. Erofeev, E., Wimmel, H.: Reachability graphs of two-transition Petri nets. In: Proceedings of the International Workshop on Algorithms & Theories for the Analysis of Event Data, Zaragoza, Spain, pp. 39–54 (2017)
26. Hujsa, T.: Contribution to the study of weighted Petri nets. Ph.D. thesis, Pierre and Marie Curie University, Paris, France (2014)

27. Hujsa, T., Delosme, J.M., Munier-Kordon, A.: On the reversibility of well-behaved weighted choice-free systems. In: Ciardo, G., Kindler, E. (eds.) Application and Theory of Petri Nets and Concurrency, pp. 334–353. Springer (2014)

28. Hujsa, T., Delosme, J.M., Munier-Kordon, A.: Polynomial sufficient conditions of well-behavedness and home markings in subclasses of weighted Petri nets. ACM Trans. Embed. Comput. Syst. **13**(4s), 141:1–141:25 (2014). https://doi.org/10.1145/2627349

29. Hujsa, T., Devillers, R.: On liveness and deadlockability in subclasses of weighted petri nets. In: van der Aalst, W., Best, E. (eds.) PETRI NETS 2017. LNCS, vol. 10258, pp. 267–287. Springer, Cham (2017). https://doi.org/10.1007/978-3-319-57861-3_16

30. Hujsa, T., Devillers, R.: On deadlockability, liveness and reversibility in subclasses of weighted Petri nets. Fundam. Inform. **161**(4), 383–421 (2018) https://doi.org/10.3233/FI-2018-1708

31. Karmarkar, N.: A new polynomial-time algorithm for linear programming. Combinatorica **4**(4), 373–395 (1984). https://doi.org/10.1007/BF02579150

32. Marchetti, O., Munier-Kordon, A.: A sufficient condition for the liveness of Weighted Event Graphs. Eur. J. Oper. Res. **197**(2), 532–540 (2009)

33. Murata, T.: Petri nets: properties, analysis and applications. Proc. IEEE **77**(4), 541–580 (1989)

34. Silva, M., Terue, E., Colom, J.M.: Linear algebraic and linear programming techniques for the analysis of place/transition net systems. In: Reisig, W., Rozenberg, G. (eds.) ACPN 1996. LNCS, vol. 1491, pp. 309–373. Springer, Heidelberg (1998). https://doi.org/10.1007/3-540-65306-6_19

35. Teruel, E., Chrzastowski-Wachtel, P., Colom, J.M., Silva, M.: On weighted T-systems. In: Jensen, K. (ed.) ICATPN 1992. LNCS, vol. 616, pp. 348–367. Springer, Heidelberg (1992). https://doi.org/10.1007/3-540-55676-1_20

36. Teruel, E., Colom, J.M., Silva, M.: Choice-free petri nets: a model for deterministic concurrent systems with bulk services and arrivals. IEEE Trans. Syst. Man Cybern. Part A **27**(1), 73–83 (1997). https://doi.org/10.1109/3468.553226

37. Teruel, E., Silva, M.: Structure theory of Equal Conflict systems. Theoret. Comput. Sci. **153**(1&2), 271–300 (1996)

38. Tredup, R.: Synthesis of Structurally Restricted b-bounded Petri Nets: Complexity Results. In: Filiot, E., Jungers, R., Potapov, I. (eds.) RP 2019. LNCS, vol. 11674, pp. 202–217. Springer, Cham (2019). https://doi.org/10.1007/978-3-030-30806-3_16

The Complexity of Synthesizing
nop-Equipped Boolean Petri Nets
from g-Bounded Inputs

Ronny Tredup[✉]

Universität Rostock, Institut für Informatik, Theoretische Informatik,
Albert-Einstein-Straße 22, 18059 Rostock, Germany
ronny.tredup@uni-rostock.de

Abstract. Boolean Petri nets equipped with nop allow places and transitions to be independent by being related by nop. We characterize for any fixed $g \in \mathbb{N}$ the computational complexity of synthesizing nop-equipped Boolean Petri nets from labeled directed graphs whose states have at most g incoming and at most g outgoing arcs.

1 Introduction

Boolean Petri nets are a basic model for the description of distributed and concurrent systems. These nets allow at most one token on each place p in every reachable marking. Therefore, p is considered a Boolean condition that is *true* if p is marked and *false* otherwise. A place p and a transition t of a Boolean Petri net N are related by one of the following Boolean *interactions*: *no operation* (nop), *input* (inp), *output* (out), *unconditionally set to true* (set), *unconditionally reset to false* (res), *inverting* (swap), *test if true* (used), and *test if false* (free). The relation between p and t determines which conditions p must satisfy to allow t's firing and which impact has the firing of t on p: The interaction inp (out) defines that p must be *true* (*false*) first and *false* (*true*) after t has fired. If p and t are related by free (used) then t's firing proves that p is *false* (*true*). The interaction nop says that p and t are independent, that is, neither need p to fulfill any condition nor has the firing of t any impact on p. If p and t are related by res (set) then p can be both *false* or *true* but after t's firing it is *false* (*true*). Also, the interaction swap does not require that p satisfies any particular condition to enable t. Here, the firing of t inverts p's Boolean value.

Boolean Petri nets are classified by the interactions of I that they use to relate places and transitions. More exactly, a subset $\tau \subseteq I$ is called a *type of net* and a net N is of type τ (a τ-net) if it applies at most the interactions of τ. So far, research has explicitly discussed seven Boolean Petri net types: *Elementary net systems* $\{\text{nop}, \text{inp}, \text{out}\}$ [9], *Contextual nets* $\{\text{nop}, \text{inp}, \text{out}, \text{used}, \text{free}\}$ [6], *event/condition nets* $\{\text{nop}, \text{inp}, \text{out}, \text{used}\}$ [2], *inhibitor nets* $\{\text{nop}, \text{inp}, \text{out}, \text{free}\}$ [8], *set nets* $\{\text{nop}, \text{inp}, \text{set}, \text{used}\}$ [5], *trace nets* $\{\text{nop}, \text{inp}, \text{out}, \text{set}, \text{res}, \text{used}, \text{free}\}$ [3], and *flip flop nets* $\{\text{nop}, \text{inp}, \text{out}, \text{swap}\}$ [10].

© Springer-Verlag GmbH Germany, part of Springer Nature 2021
M. Koutny et al. (Eds.): ToPNoC XV, LNCS 12530, pp. 101–125, 2021.
https://doi.org/10.1007/978-3-662-63079-2_5

However, since we have eight interactions to choose from, there are actually a total of 256 different types.

This paper addresses the computational complexity of the τ-*synthesis* problem. It consists in deciding whether a given directed labeled graph A, also called *transition system*, is isomorphic to the reachability graph of a τ-net N and in constructing N if it exists. It has been shown that τ-*synthesis* is NP-complete if $\tau = \{\mathsf{nop}, \mathsf{inp}, \mathsf{out}\}$ [1], even if the inputs are strongly restricted [14,17]. On the contrary, in [10], it has been shown that it becomes polynomial if $\tau = \{\mathsf{nop}, \mathsf{inp}, \mathsf{out}, \mathsf{swap}\}$. These opposing results motivate the question which interactions of I make the synthesis problem hard and which make it tractable. In our previous work of [13,15,16], we answer this question partly and reveal the computational complexity of 120 of the 128 types that allow nop.

In this paper, we investigate for fixed $g \in \mathbb{N}$ the computational complexity of τ-synthesis restricted to g-bounded inputs, that is, every state of A has at most g incoming and at most g outgoing arcs. On the one hand, inputs of practical applications tend to have a low bound g such as benchmarks of digital hardware design [4]. On the other hand, considering restricted inputs hopefully gives us a better understanding of the problem's hardness. Thus, g-bounded inputs are interesting from both the practical and the theoretical point of view. In this paper, we completely characterize the complexity of τ-synthesis restricted to g-bounded inputs for all types that allow places and transitions to be independent, that is, which contain nop. Figure 1 summarizes our findings: For the types of §1 and §2, we showed hardness of synthesis without restriction in [15]. In this paper, we strengthen these results to 2- and 3-bounded inputs, respectively, and show that these bounds are tight. The hardness result of the types of §3 originates from [16]. This paper shows that a bound less than 2 makes synthesis tractable. Hardness for the types of §4 to §8 has been shown for 2-bounded inputs in [16]. In this paper, we strengthen this results to 1-bounded inputs. The hardness part for the types of §9 origin from [13]. In this paper, we argue that the bound 2 is tight. Finally, while the results of §10 are new, the ones of §11 have been found in [15].

For all considered types τ, the corresponding hardness results are based on a reduction of the so-called *cubic monotone one-in-three 3SAT* problem [7]. All reductions follow a common approach that represents clauses by directed labeled paths. Thus, this paper also contributes a very general way to prove NP-completeness of synthesis of Boolean types of nets.

2   Preliminaries

Transition Systems. A *transition system* (TS) $A = (S, E, \delta)$ is a directed labeled graph with states S, events E and partial *transition function* $\delta : S \times E \longrightarrow S$, where $\delta(s, e) = s'$ is interpreted as $s \xrightarrow{e} s'$. For $s \xrightarrow{e} s'$ we say s is a source and s' is a sink of e, respectively. An event e *occurs* at a state s, denoted by $s \xrightarrow{e}$, if $\delta(s, e)$ is defined. An *initialized* TS $A = (S, E, \delta, s_0)$ is a TS with a distinct state $s_0 \in S$ where every state $s \in S$ is *reachable* from s_0 by a directed labeled path. TSs in this paper are *deterministic* by design as their state

§	Type of net τ	g	Complexity	#
1	{nop, inp, free}, {nop, inp, used, free}, {nop, out, used}, {nop, out, used, free}	≥ 2	NP-complete	4
		< 2	polynomial	
2	{nop, set, res} $\cup\,\omega$ and $\emptyset \neq \omega \subseteq$ {used, free}	≥ 3	NP-complete	3
		< 3	polynomial	
3	{nop, inp, set}, {nop, inp, set, used}, {nop, inp, res, set} $\cup\,\omega$ and $\omega \subseteq$ {out, used, free}, {nop, out, res}, {nop, out, res, free}, {nop, out, res, set} $\cup\,\omega$ and $\omega \subseteq$ {inp, used, free}	≥ 2	NP-complete	16
		< 2	polynomial	
4	{nop, inp, out, set} $\cup\,\omega$ or {nop, inp, out, res} $\cup\,\omega$ and $\omega \subseteq$ {used, free}	≥ 1	NP-complete	8
5	{nop, inp, set, free}, {nop, inp, set, used, free}, {nop, out, res, used}, {nop, out, res, used, free}	≥ 1	NP-complete	4
6	{nop, inp, res, swap} $\cup\,\omega$ or {nop, out, set, swap} $\cup\,\omega$ and $\omega \subseteq$ {used, free}	≥ 1	NP-complete	8
7	{nop, inp, set, swap} $\cup\,\omega$ and $\omega \subseteq$ {out, res, used, free}, {nop, out, res, swap} $\cup\,\omega$ and $\omega \subseteq$ {inp, set, used, free}	≥ 1	NP-complete	28
8	{nop, inp, out} $\cup\,\omega$ and $\omega \subseteq$ {used, free}	≥ 1	NP-complete	4
9	{nop, set, swap} $\cup\,\omega$, {nop, res, swap} $\cup\,\omega$, {nop, res, set, swap} $\cup\,\omega$ and $\emptyset \neq \omega \subseteq$ {used, free}	≥ 2	NP-complete	9
		< 2	polynomial	
10	{nop, inp}, {nop, inp, used}, {nop, out}, {nop, out, free} {nop, set, swap}, {nop, res, swap}, {nop, set, res}, {nop, set, res, swap}	≥ 0	polynomial	8
11	{nop, res} $\cup\,\omega$ and $\omega \subseteq$ {inp, used, free}, {nop, set} $\cup\,\omega$ and $\omega \subseteq$ {out, used, free}, {nop, swap} $\cup\,\omega$ and $\omega \subseteq$ {inp, out, used, free}, {nop} $\cup\,\omega$ and $\omega \subseteq$ {used, free}	≥ 0	polynomial	36

Fig. 1. The computational complexity of Boolean net synthesis from g-bounded TS for all types that contain nop.

transition behavior is given by a (partial) function. Let $g \in \mathbb{N}$. An initialized TS A is called g-bounded if for all $s \in S(A)$ the number of incoming and outgoing arcs at s is restricted by g: $|\{e \in E(A) \mid \overset{e}{\longrightarrow}s\}| \leq g$ and $|\{e \in E(A) \mid s\overset{e}{\longrightarrow}\}| \leq g$.

Boolean Types of Nets [2]. The following notion of Boolean types of nets serves as vehicle to capture many Boolean Petri nets in a uniform way. A *Boolean type of net* $\tau = (\{0, 1\}, E_\tau, \delta_\tau)$ is a TS such that E_τ is a subset of the Boolean interactions: $E_\tau \subseteq I = $ {nop, inp, out, set, res, swap, used, free}. The interactions $i \in I$ are binary partial functions $i : \{0, 1\} \rightarrow \{0, 1\}$ as defined in Fig. 2. For all $x \in \{0, 1\}$ and all $i \in E_\tau$ the transition function of τ is defined by $\delta_\tau(x, i) = i(x)$. Notice that I contains all binary partial functions $\{0, 1\} \rightarrow \{0, 1\}$ except for the entirely undefined function $\perp$. Even if a type τ includes $\perp$, this event can never occur, so it would be useless. Thus, I is complete for deterministic Boolean types of nets, and that means there are a total of 256 of them. By definition, a Boolean type τ is completely determined by its event set E_τ. Hence, in the following we identify τ with E_τ, cf. Fig. 3. Moreover, for readability, we group interactions by enter = {out, set, swap}, exit = {inp, res, swap}, keep$^+$ = {nop, set, used}, and keep$^-$ = {nop, res, free}.

x	nop(x)	inp(x)	out(x)	set(x)	res(x)	swap(x)	used(x)	free(x)
0	0		1	1	0	1		0
1	1	0		1	0	0	1	

Fig. 2. All interactions in I. An empty cell means that the column's function is undefined on the respective x. The entirely undefined function is missing in I.

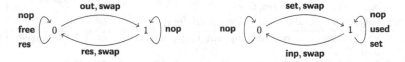

Fig. 3. Left: $\tau = \{$nop, out, res, swap, free$\}$. Right: $\tilde{\tau} = \{$nop, inp, set, swap, used$\}$. τ and $\tilde{\tau}$ are isomorphic. The isomorphism $\phi : \tau \rightarrow \tilde{\tau}$ is given by $\phi(s) = 1 - s$ for $s \in \{0,1\}$, $\phi(i) = i$ for $i \in \{$nop, swap$\}$, $\phi(\text{out}) = \text{inp}$, $\phi(\text{res}) = \text{set}$ and $\phi(\text{free}) = \text{used}$.

τ-Nets. Let $\tau \subseteq I$. A Boolean Petri net $N = (P, T, H_0, f)$ of type τ, (τ-net) is given by finite and disjoint sets P of places and T of transitions, an initial marking $H_0 : P \longrightarrow \{0,1\}$, and a (total) flow function $f : P \times T \rightarrow \tau$. A τ-net realizes a certain behavior by firing sequences of transitions: A transition $t \in T$ can fire in a marking $M : P \longrightarrow \{0,1\}$ if $\delta_\tau(M(p), f(p,t))$ is defined for all $p \in P$. By firing, t produces the next marking $M' : P \longrightarrow \{0,1\}$ where $M'(p) = \delta_\tau(M(p), f(p,t))$ for all $p \in P$. This is denoted by $M \xrightarrow{t} M'$. Given a τ-net $N = (P, T, H_0, f)$, its behavior is captured by a transition system A_N, called the reachability graph of N. The state set of A_N consists of all markings that, starting from initial state H_0, can be reached by firing a sequence of transitions. For every reachable marking M and transition $t \in T$ with $M \xrightarrow{t} M'$ the state transition function δ of A is defined as $\delta(M, t) = M'$.

τ-Regions. Let $\tau \subseteq I$. If an input A of τ-synthesis allows a positive decision then we want to construct a corresponding τ-net N purely from A. Since A and A_N are isomorphic, N's transitions correspond to A's events. However, the notion of a place is unknown for TSs. So-called regions mimic places of nets: A τ-region of a given $A = (S, E, \delta, s_0)$ is a pair (sup, sig) of $support$ $sup : S \rightarrow S_\tau = \{0,1\}$ and $signature$ $sig : E \rightarrow E_\tau = \tau$ where every transition $s \xrightarrow{e} s'$ of A leads to a transition $sup(s) \xrightarrow{sig(e)} sup(s')$ of τ. While a region divides S into the two sets $sup^{-1}(b) = \{s \in S \mid sup(s) = b\}$ for $b \in \{0,1\}$, the events are cumulated by $sig^{-1}(i) = \{e \in E \mid sig(e) = i\}$ for all available interactions $i \in \tau$. We also use $sig^{-1}(\tau') = \{e \in E \mid sig(e) \in \tau'\}$ for $\tau' \subseteq \tau$. A region (sup, sig) models a place p and the corresponding part of the flow function f. In particular, $sig(e)$ models $f(p, e)$ and $sup(s)$ models $M(p)$ in the marking $M \in RS(N)$ corresponding to $s \in S(A)$. Every set $\mathcal{R}$ of τ-regions of A defines the $synthesized$ τ-net $N_A^{\mathcal{R}} = (\mathcal{R}, E, f, H_0)$ with flow function $f((sup, sig), e) = sig(e)$ and initial marking $H_0((sup, sig)) = sup(s_0)$ for all $(sup, sig) \in \mathcal{R}, e \in E$. It is well known that $A_{N_A^{\mathcal{R}}}$ and A are isomorphic if and only if $\mathcal{R}$'s regions solve certain separation atoms [2], to be introduced next. A pair (s, s') of distinct states of A defines a *state separation atom* (SSP atom). A τ-region $R = (sup, sig)$ solves (s, s') if $sup(s) \neq sup(s')$. The meaning of R is to ensure that $N_A^{\mathcal{R}}$ contains at least one place R such that $M(R) \neq M'(R)$ for the markings M and M' corresponding to s and s', respectively. If there is a τ-region that solves (s, s') then s and s' are called τ-*solvable*. If every SSP atom of A is τ-solvable then A has the τ-*state separation property* (τ-SSP). A pair (e, s) of event $e \in E$ and state $s \in S$ where e does not occur at s, that is $\neg s \overset{e}{\longrightarrow}$, defines an *event state separation atom* (ESSP atom). A τ-region $R = (sup, sig)$ *solves* (e, s) if $sig(e)$ is not defined on $sup(s)$ in τ, that is, $\neg \delta_\tau(sup(s), sig(e))$. The meaning of R is to ensure that there is at least one place R in $N_A^{\mathcal{R}}$ such that $\neg M \overset{e}{\longrightarrow}$ for the marking M corresponding to s. If there is a τ-region that solves (e, s) then e and s are called τ-*solvable*. If every ESSP atom of A is τ-solvable then A has the τ-*event state separation property* (τ-ESSP). A set $\mathcal{R}$ of τ-regions of A is called τ-*admissible* if for every of A's (E)SSP atoms there is a τ-region R in $\mathcal{R}$ that solves it. The following lemma, borrowed from [2, p.163], summarizes the already implied connection between the existence of τ-admissible sets of A and (the solvability of) τ-synthesis:

Lemma 1 ([2]). *A TS A is isomorphic to the reachability graph of a τ-net N if and only if there is a τ-admissible set $\mathcal{R}$ of A such that $N = N_A^{\mathcal{R}}$.*

We say a τ-net N τ-*solves* A if A_N and A are isomorphic. By Lemma 1, deciding if A is τ-solvable reduces to deciding whether it has the τ-(E)SSP. Moreover, it is easy to see that if τ and $\tilde{\tau}$ are isomorphic then deciding the τ-(E)SSP reduces to deciding the $\tilde{\tau}$-(E)SSP:

Lemma 2 (Without proof). *If τ and $\tilde{\tau}$ are isomorphic types of nets then a TS A has the τ-(E)SSP if and only if A has the $\tilde{\tau}$-(E)SSP.*

In particular, we benefit from the isomorphisms that map nop to nop, swap to swap, inp to out, set to res, used to free, and vice versa.

$$s_0 \overset{a}{\longrightarrow} s_1 \overset{a}{\longleftarrow} s_2 \qquad s_0 \overset{a}{\longrightarrow} s_1 \overset{a}{\longrightarrow} s_2 \qquad s_0 \overset{a}{\longleftarrow} s_1 \overset{a}{\longleftarrow} s_2 \qquad s_0 \overset{a}{\longrightarrow} s_1 \overset{a}{\longrightarrow} s_2 \overset{a}{\longrightarrow} s_3$$

TS A_1.   TS A_2.   TS A_3.   TS A_4.

Fig. 4. Let $\tau = \{\mathsf{nop}, \mathsf{set}, \mathsf{swap}, \mathsf{free}\}$. The TSs $A_1, \ldots, A_4$ give examples for the presence and absence of the τ-(E)SSP: TS A_1 has the τ-ESSP as a occurs at every state. It has also the τ-SSP: The region $R = (sup, sig)$ where $sup(s_0) = sup(s_2) = 1$, $sup(s_1) = 0$ and $sig(a) = \mathsf{swap}$ separates the pairs s_0, s_1 and s_2, s_1. Moreover, the region $R' = (sup', sig')$ where $sup'(s_0) = 0$ and $sup'(s_1) = sup'(s_2) = 1$ and $sig'(a) = \mathsf{set}$ separates s_0 and s_1. Notice that R and R' can be translated into $\tilde{\tau}$-regions, where $\tilde{\tau} = \{\mathsf{nop}, \mathsf{res}, \mathsf{swap}, \mathsf{used}\}$, via the isomorphism of Fig. 3. For example, if $s \in S(A_1)$ and $e \in E(A_1)$ and $sup''(s) = \phi(sup(s))$ and $sig''(e) = \phi(sig(e))$ then the resulting $\tilde{\tau}$-region $R'' = (sup'', sig'')$ separates s_0, s_1 and s_2, s_1. Thus, A_1 has also $\tilde{\tau}$-(E)SSP. TS A_2 has the τ-SSP but not the τ-ESSP as event a is not inhibitable at the state s_2. TS A_3 has the τ-ESSP (a occurs at every state) but not the τ-SSP as s_1 and s_2 are not separable. TS A_4 has neither the τ-ESSP nor the τ-SSP.

3   Hardness Results

In this section, for several types of nets $\tau \subseteq I$ and fixed $g \in \mathbb{N}$, we show that τ-synthesis is NP-complete even if the input TS A is g-bounded, cf. Fig. 1. Since τ-synthesis is known to be in NP for all $\tau \subseteq I$ [16], we restrict ourselves to the hardness part. All proofs are based on a reduction of the problem *cubic monotone one-in-three* 3-*SAT* which has been shown to be NP-complete in [7]. The input for this problem is a Boolean expression $\varphi = \{\zeta_0, \ldots, \zeta_{m-1}\}$ of m negation-free three-clauses $\zeta_i = \{X_{i_0}, X_{i_1}, X_{i_2}\}$ such that every variable $X \in V(\varphi) = \bigcup_{i=0}^{m-1} \zeta_i$ occurs in exactly three clauses. Notice that the latter implies $|V(\varphi)| = m$. Moreover, we assume without loss of generality that if $\zeta_i = \{X_{i_0}, X_{i_1}, X_{i_2}\}$ then $i_0 < i_1 < i_2$. The question to answer is whether there is a subset $M \subseteq V(\varphi)$ with $|M \cap \zeta_i| = 1$ for all $i \in \{0, \ldots, m-1\}$. For all considered types of nets τ and corresponding bounds g, we reduce a given instance φ to a g-bounded TS A_φ^τ such that the following two conditions are true: Firstly, the TS A_φ^τ has an ESSP atom α which is τ-solvable if and only if there is a one-in-three model M of φ. Secondly, if the ESSP atom α is τ-solvable then all ESSP and SSP atoms of A_φ^τ are also τ-solvable. A reduction that satisfies these conditions proves the NP-hardness of τ-synthesis as follows: If φ has a one-three-model then the conditions ensure that the TS A_φ^τ has the τ-(E)SSP and thus is τ-solvable. Conversely, if A_φ^τ is τ-solvable then, by definition, it has the τ-ESSP. In particular, there is a τ-region that solves α which, by the first condition, implies that φ has a one-in-three model. Consequently, A_φ^τ is τ-solvable if and only if φ has a one-in-three model. Due to space restrictions, we omit for all considered types the proof that A_φ^τ satisfies the second condition, that is, that the solvability of α implies the (E)SSP. However, the corresponding proofs can be found in the technical report [11].

A key idea, applied by all reductions in one way or another, is the representation of every clause $\zeta_i = \{X_{i_0}, X_{i_1}, X_{i_2}\}$, $i \in \{0, \ldots, m-1\}$, by a directed labeled path of A_φ^τ on which the variables of ζ_i occur as events:

$$s_{i,0} \ldots s_{i,j} \xrightarrow{X_{i_0}} s_{i,j+1} \ldots s_{i,j'} \xrightarrow{X_{i_1}} s_{i,j'+1} \ldots s_{i,j''} \xrightarrow{X_{i_2}} s_{i,j''+1} \ldots s_{i,n}$$

The reductions ensure that if a τ-region (sup, sig) solves the atom α then $sup(s_{i,0}) \neq sup(s_{i,n})$. This makes the image of this path under (sup, sig) a directed path from 0 to 1 or from 1 to 0 in τ. Thus, there has to be an event e on the path that causes the state change from $sup(s_{i,0})$ to $sup(s_{i,n})$ by $sig(e)$. We exploit this property and ensure that this state change is unambiguously done by (the signature of) exactly one variable event per clause. As a result, the corresponding variable events define a searched model of φ via their signature. The proof of the following theorem gives a first example of this approach, and Fig. 5 shows a full example reduction.

Theorem 1. *For any fixed $g \geq 2$, deciding if a g-bounded TS A is τ-solvable is NP-complete if $\tau = \{nop, inp, free\}$, $\tau = \{nop, inp, used, free\}$, $\tau = \{nop, out, used\}$ and $\tau = \{nop, out, used, free\}$.*

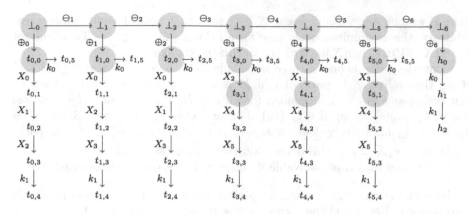

Fig. 5. The TS A_φ^τ for $\varphi = \{\zeta_0, \dots, \zeta_5\}$ with clauses $\zeta_0 = \{X_0, X_1, X_2\}$, $\zeta_1 = \{X_0, X_2, X_3\}$, $\zeta_2 = \{X_0, X_1, X_3\}$, $\zeta_3 = \{X_2, X_4, X_5\}$, $\zeta_4 = \{X_1, X_4, X_5\}$, $\zeta_5 = \{X_3, X_4, X_5\}$. The red colored area sketc.hes the τ-region (sup, sig) that solves (k_1, h_0) and corresponds to the one-in-three model $M = \{X_0, X_4\}$. (Color figure online)

Proof. We argue for $\tau \in \{\{\mathsf{nop}, \mathsf{inp}, \mathsf{free}\}, \{\mathsf{nop}, \mathsf{inp}, \mathsf{used}, \mathsf{free}\}\}$, which by Lemma 2 proves the claim for the other types, too.

Firstly, the TS A_φ^τ has the following gadget H (left hand side) that provides the events k_0, k_1 and the atom $\alpha = (k_1, h_0)$. Secondly, it has for every clause $\zeta_i = \{X_{i_0}, X_{i_1}, X_{i_2}\}$ the following gadget T_i (right hand side) that applies k_0, k_1 and $\zeta_i's$ variables as events.

$$h_0 \xrightarrow{k_0} h_1 \xrightarrow{k_1} h_2 \qquad t_{i,0} \xrightarrow{X_{i_0}} t_{i,1} \xrightarrow{X_{i_1}} t_{i,2} \xrightarrow{X_{i_2}} t_{i,3} \xrightarrow{k_1} t_{i,4}$$
$$\downarrow{k_0}$$
$$t_{i,5}$$

Finally, A_φ^τ uses the states $\perp_0, \dots, \perp_m$ and events $\ominus_1, \cdots \ominus_m$ and $\oplus_0, \dots, \oplus_m$ to join the gadgets $T_0, \dots, T_{m-1}$ and H by $\perp_i \xrightarrow{\ominus_{i+1}} \perp_{i+1}$ and $\perp_i \xrightarrow{\oplus_i} t_{i,0}$, for all $i \in \{0, \dots, m-1\}$, and $\perp_m \xrightarrow{\oplus_m} h_0$, cf. Fig. 5.

The gadget H ensures that if (sup, sig) is a region that solves α then $sup(h_0) = 1$ and $sig(k_1) = \mathsf{free}$ which implies $sup(h_1) = 0$ and $sig(k_0) = \mathsf{inp}$. This is because $sig(k_1) \in \{\mathsf{inp}, \mathsf{used}\}$ and $sup(h_0) = 0$ implies $sig(k_0) \in \{\mathsf{out}, \mathsf{set}, \mathsf{swap}\}$, which is impossible. Consequently, $s \xrightarrow{k_0}$ and $s' \xrightarrow{k_1}$ imply $sup(s) = 1$ and $sup(s') = 0$, respectively. The TS A_φ^τ uses these properties to ensure via $T_0, \dots, T_{m-1}$ that the region (sup, sig) implies a one-in-three model of φ.

More exactly, if $i \in \{0, \dots, m-1\}$ then for T_i we have by $t_{i,0} \xrightarrow{k_0}$ and $t_{i,3} \xrightarrow{k_1}$ that $sup(t_{i,0}) = 1$ and $sup(t_{i,3}) = 0$. Thus, there is an event X_{i_j}, where $j \in \{0, 1, 2\}$, such that $sig(X_{i_j}) = \mathsf{inp}$. Clearly, if $sig(X_{i_j}) = \mathsf{inp}$ then $sig(X_{i_\ell}) \neq \mathsf{inp}$ for all $j < \ell \in \{0, 1, 2\}$ as X_{i_ℓ}'s sources have a 0-support. Consequently, there is *exactly one* variable event $X \in \zeta_i$ such that $sig(X) = $ inp. Since i was arbitrary, this is simultaneously true for all clauses $\zeta_0, \ldots, \zeta_{m-1}$. Thus, the set $M = \{X \in V(\varphi) \mid sig(X) = \text{inp}\}$ is a one-in-three model of φ.

Conversely, if φ is one-in-three satisfiable then there is a τ-region (sup, sig) of A_φ^τ that solves α. In particular, if M is a one-in-three model of φ then we first define $sup(\bot_0) = 1$. Secondly, for all $e \in E(A_\varphi^\tau)$ we define $sig(e) = $ free if $e = k_1$, $sig(e) = $ inp if $e \in \{k_0\} \cup M$ and else $sig(e) = $ nop. Since A_φ^τ is reachable, by inductively defining $sup(s_{i+1}) = \delta_\tau(sup(s_i), sig(e_i))$ for all paths $\bot_0 \xrightarrow{e_1} s_1 \ldots s_{n-1} \xrightarrow{e_n} s_n$, this defines a fitting region (sup, sig), cf. Fig. 5.

This proves that α is τ-solvable if and only if φ is one-in-three satisfiable.

In the remainder of this section, we present the remaining hardness results in accordance to Fig. 1 and the corresponding reductions that prove them.

Theorem 2. *For any fixed $g \geq 3$, deciding if a g-bounded TS A is τ-solvable is NP-complete if $\tau = \{$nop, set, res$\} \cup \omega$ and $\emptyset \neq \omega \subseteq \{$used, free$\}$.*

Proof. The TS A_φ^τ has the following gadgets H_0, H_1 and H_2 (in this order):

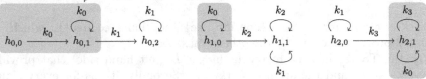

The gadget H_0 provides $\alpha = (k_0, h_{0,2})$. By symmetry, A_I^τ is $\{$nop, set, res, used$\}$-solvable if and only if it is $\{$nop, set, res, free$\}$- or $\{$nop, set, res, free, used$\}$-solvable. Thus, in the following we assume $\tau = \{$nop, set, res, used$\}$, $sig(k_0) = $ used and $sup(h_{0,2}) = 0$ if (sup, sig) τ-solves α. As a result, by $sig(k_0) = $ used, implying $sup(h_{0,1}) = 1$, and $sup(h_{0,2}) = 0$ we have $sig(k_1) = $ res. Especially, if $\xrightarrow{k_0} s$ then $sup(s) = 1$ and if $\xrightarrow{k_1} s$ then $sup(s) = 0$. Thus, $sup(h_{1,0}) = sup(h_{2,1}) = 1$ and $sup(h_{1,1}) = sup(h_{2,0}) = 0$ which implies $sig(k_2) = $ res and $sig(k_3) = $ set.

The construction uses k_2 and k_3 to produce some neutral events. More exactly, the TS A_φ^τ implements for all $j \in \{0, \ldots, 16m - 1\}$ the following gadget F_j that uses k_2 and k_3 to ensure that the events z_j are neutral:

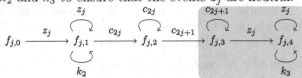

By $sig(k_2) = $ res and $sig(k_3) = $ set we have $sup(f_{j,1}) = 0$ and $sup(f_{j,4}) = 1$. This implies $\xrightarrow{sig(z_j)} 0$ and $\xrightarrow{sig(z_j)} 1$ and thus $sig(z_j) = $ nop.

Finally, for every $i \in \{0, \ldots, m-1\}$ and clause $\zeta_i = \{X_{i_0}, X_{i_1}, X_{i_2}\}$, the TS A_φ^τ has the following four gadgets $T_{i,0}, T_{i,1} T_{i,2}$ and $T_{i,3}$ (in this order):

$$t_{i,0,0} \xrightarrow{z_{16i}} t_{i,0,1} \xrightarrow{X_{i_0}} t_{i,0,2} \xrightarrow{z_{16i+1}} t_{i,0,3} \xrightarrow{X_{i_1}} t_{i,0,4} \xrightarrow{z_{16i+2}} t_{i,0,5} \xrightarrow{X_{i_2}} t_{i,0,6} \xrightarrow{z_{16i+3}} t_{i,0,7}$$

loops: $z_{16i},\ X_{i_0},\ z_{16i+1},\ X_{i_1},\ z_{16i+2},\ X_{i_2},\ z_{16i+3}$; $k_0,\ y_{3i+1},\ y_{3i+2},\ k_1$

$$t_{i,1,0} \xrightarrow{z_{16i+4}} t_{i,1,1} \xrightarrow{X_{i_1}} t_{i,1,2} \xrightarrow{z_{16i+5}} t_{i,1,3} \xrightarrow{X_{i_2}} t_{i,1,4} \xrightarrow{z_{16i+6}} t_{i,1,5} \xrightarrow{X_{i_0}} t_{i,1,6} \xrightarrow{z_{16i+7}} t_{i,1,7}$$

loops: $z_{16i+4},\ X_{i_1},\ z_{16i+5},\ X_{i_2},\ z_{16i+6},\ X_{i_0},\ z_{16i+7}$; $k_0,\ y_{3i},\ y_{3i},\ k_1$

$$t_{i,2,0} \xrightarrow{z_{16i+8}} t_{i,2,1} \xrightarrow{X_{i_2}} t_{i,2,2} \xrightarrow{z_{16i+9}} t_{i,2,3} \xrightarrow{X_{i_0}} t_{i,2,4} \xrightarrow{z_{16i+10}} t_{i,2,5} \xrightarrow{X_{i_1}} t_{i,2,6} \xrightarrow{z_{16i+11}} t_{i,2,7}$$

loops: $z_{16i+8},\ X_{i_2},\ z_{16i+9},\ X_{i_0},\ z_{16i+10},\ X_{i_1},\ z_{16i+11}$; $k_0,\ y_{3i+1},\ k_1$

$$t_{i,3,0} \xrightarrow{z_{16i+12}} t_{i,3,1} \xrightarrow{y_{3i}} t_{i,3,2} \xrightarrow{z_{16i+13}} t_{i,3,3} \xrightarrow{y_{3i+1}} t_{i,3,4} \xrightarrow{z_{16i+14}} t_{i,3,5} \xrightarrow{y_{3i+2}} t_{i,3,6} \xrightarrow{z_{16i+15}} t_{i,3,7}$$

loops: $z_{16i+12},\ y_{3i},\ z_{16i+13},\ y_{3i+1},\ z_{16i+14},\ y_{3i+2},\ z_{16i+15}$; $k_1,\ k_0$

$T_{i,0}, \ldots, T_{i,4}$ ensure that there is exactly one $X \in \zeta_i$ with $sig(X) = \mathsf{res}$: By $sig(k_0) = \mathsf{used}$ and $sig(k_1) = \mathsf{res}$ we get $sup(t_{i,0,0}) = sup(t_{i,1,0}) = sup(t_{i,2,0}) = sup(t_{i,3,7}) = 1$ and $sup(t_{i,0,7}) = sup(t_{i,1,7}) = sup(t_{i,2,7}) = sup(t_{i,3,0}) = 0$. Since $z_{16i}, \ldots, z_{16i+11}$ are neutral, this implies $sup(t_{i,0,6}) = sup(t_{i,1,6}) = sup(t_{i,2,6}) = 0$ and that there is a variable event with a res-signature. Moreover, by $sup(t_{i,3,0}) = 0$ and $sup(t_{i,3,7}) = 1$ and the neutrality of $z_{16i+12}, \ldots, z_{16i+15}$ there is an event of $y_{3i}, y_{3i+1}, y_{3i+2}$ with a set-signature. We argue that there is exactly one variable event with a res-signature: By $sup(t_{i,0,6}) = sup(t_{i,1,6}) = sup(t_{i,2,6}) = 0$, we have $sig(X) \notin \{\mathsf{set}, \mathsf{used}\}$ for all $X \in \{X_{i_0}, X_{i_1}, X_{i_2}\}$. Hence, if $sig(X_{i_0}) = \mathsf{res}$ then $sup(t_{i,0,2}) = \cdots = sup(t_{i,0,6}) = 0$ which implies $sig(y_{3i+1}) \neq \mathsf{set}$ and $sig(y_{3i+2}) \neq \mathsf{set}$ and thus $sig(y_{3i}) = \mathsf{set}$. By $sig(y_{3i}) = \mathsf{set}$ we have $sup(t_{i,1,2}) = sup(t_{i,1,4}) = 1$ which implies $sig(X_{i_1}) \neq \mathsf{res}$ and $sig(X_{i_2}) \neq \mathsf{res}$.

If $sig(X_{i_1}) = \mathsf{res}$, then $sup(t_{i,0,4}) = sup(t_{i,1,2}) = 0$ which implies $sig(y_{3i}) \neq \mathsf{set}$ and $sig(y_{3i+2}) \neq \mathsf{set}$ and thus $sig(y_{3i+1}) = \mathsf{set}$. By $sig(y_{3i+1}) = \mathsf{set}$ we have $sup(t_{i,0,2}) = sup(t_{i,2,2}) = 1$ which implies $sig(X_{i_0}) \neq \mathsf{res}$ and $sig(X_{i_2}) \neq \mathsf{res}$.

Since $sig(X_{i_0}) = \mathsf{res}$ or $sig(X_{i_1}) = \mathsf{res}$ implies $sig(X_{i_2}) \neq \mathsf{res}$, we conclude that $sig(X_{i_2}) = \mathsf{res}$ implies $sig(X_{i_0}) \neq \mathsf{res}$ and $sig(X_{i_1}) \neq \mathsf{res}$. Thus, there is exactly one variable of the i-th clause with a signature res. Hence, the set $M = \{X \in V(\varphi) \mid sig(X) = \mathsf{res}\}$ is a one-in-three model of φ.

To finally build A_φ^τ, we use the states $\bot = \{\bot_0, \ldots, \bot_{20m+2}\}$ and the events $\oplus = \{\oplus_0, \ldots, \oplus_{20m+2}\}$ and $\ominus = \{\ominus_1, \ldots, \ominus_{20m+2}\}$. The states of $\bot$ are connected by $\bot_j \xrightarrow{\ominus_{j+1}} \bot_{j+1}$ and $\bot_{j+1} \xrightarrow{\ominus_{j+1}} \bot_{j+1}$ for $j \in \{0, \ldots, 20m + 1\}$. Let $x = 16m + 3$ and $y = 19m + 3$. For all $i \in \{0, 1, 2\}$, for all $\ell \in \{0, \ldots, 16m - 1\}$ and for all $j \in \{0, \ldots, m\}$ we add the following edges that connect the gadgets

H_0, H_1, H_2 and $F_0, \ldots, F_{16m-1}$ and $T_{0,0}, T_{0,1}, T_{0,2}, \ldots, T_{m-1,0}, T_{m-1,1} T_{m-1,2}$ and

$$\bot_i \xrightarrow{\oplus_i} h_{i,0} \quad\circlearrowright^{\oplus_i} \qquad \bot_{\ell+3} \xrightarrow{\oplus_{\ell+3}} f_{\ell,0} \quad\circlearrowright^{\oplus_{\ell+3}} \qquad \bot_{x+3j+i} \xrightarrow{\oplus_{x+3j+i}} t_{j,i,0} \quad\circlearrowright^{\oplus_{x+3j+i}} \qquad \bot_{y+j} \xrightarrow{\oplus_{y+j}} t_{j,3,0} \quad\circlearrowright^{\oplus_{y+j}}$$

If M is a one-in-three model of φ then α is τ-solvable by a τ-region (sup, sig): If $s \in \{h_{0,0}, h_{1,0}, h_{2,1}\}$ or $\{f_{j,0} \mid j \in \{0, \ldots, 16m-1\}\}$ then $sup(s) = 1$. The support values of the states of $T_{i,0}, \ldots, T_{i,3}$, where $i \in \{0, \ldots, m-1\}$, are defined in accordance to which event of $X_{i_0}, X_{i_1}, X_{i_2}$ belongs to M. The red colored area above sketches $X_{i_0} \in M$. Moreover, we define $sup(s) = 0$ for all $s \in \bot$. Let $e \in E(A_\varphi^\tau) \setminus \oplus$. We define $sig(e) = $ used if $e = k_0$ and $sig(e) = $ res if $e \in \{k_1\} \cup M$. For all $i \in \{0, \ldots, m-1\}$ and clauses $\{X_{i_0}, X_{i_1}, X_{i_2}\}$ and all $j \in \{0, 1, 2\}$ we set $sig(e) = $ set if $e = y_{3i+j}$ and $X_{i_j} \in M$. Otherwise, we define $sig(e) = $ nop. For all events $e \in \oplus$ and edges $s \xrightarrow{e} s'$ of A we define $sig(e) = $ set if $sup(s') = 1$ and, otherwise, $sig(e) = $ nop. The resulting τ-region (sup, sig) of A_φ^τ solves α.   □

Theorem 3. *For any fixed $g \geq 2$, deciding if a g-bounded TS A is τ-solvable is NP-complete if (1) $\tau = \{\text{nop}, \text{inp}, \text{set}\}$ or $\tau = \{\text{nop}, \text{inp}, \text{set}, \text{used}\}$ or $\tau = \{\text{nop}, \text{inp}, \text{res}, \text{set}\} \cup \omega$ and $\omega \subseteq \{\text{out}, \text{used}, \text{free}\}$ or if (2) $\tau = \{\text{nop}, \text{out}, \text{res}\}$ or $\tau = \{\text{nop}, \text{out}, \text{res}, \text{free}\}$ or $\tau = \{\text{nop}, \text{out}, \text{res}, \text{set}\} \cup \omega$ and $\omega \subseteq \{\text{inp}, \text{used}, \text{free}\}$.*

Proof. We present a reduction for the types of (1). By Lemma 2, this proves the claim also for the types of (2). The TS A_φ^τ has the following gadget H:

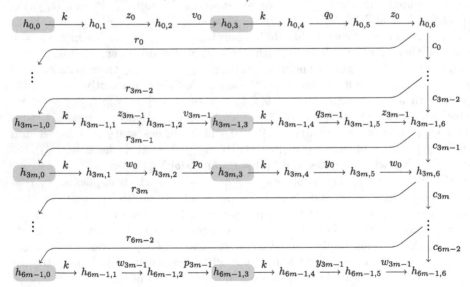

The intention of the gadget H is to provide the atom $\alpha = (k, h_{0,6})$ and the events of $Z = \{z_0, \ldots, z_{3m-1}\}$, $V = \{v_0, \ldots, v_{3m-1}\}$ and $W = \{w_0, \ldots, w_{3m-1}\}$.

Moreover, the TS A_φ^τ has the following two gadgets F_0 and F_1 and for all $i \in \{0, \ldots, 6m - 2\}$ the following gadget G_i (in this order):

$$f_{0,0} \xrightarrow{k} f_{0,1} \xrightarrow{n} f_{0,2} \xrightarrow{z_0} f_{0,3} \xrightarrow{k} f_{0,4} \qquad f_{1,0} \xrightarrow{q_0} f_{1,1} \xrightarrow{k} f_{1,2}$$

$$\begin{array}{ccc} g_{i,0} & \xrightarrow{c_i} & g_{i,1} \\ {\scriptstyle k}\downarrow & & \downarrow{\scriptstyle k} \\ g_{i,2} & \xrightarrow[c_i]{} & g_{i,3} \end{array}$$

Finally, the TS A_φ^τ has for every clause $\zeta_i = \{X_{i_0}, X_{i_1}, X_{i_2}\}$, $i \in \{0, \ldots, m - 1\}$, the following gadgets $T_{i,0}, T_{i,1}$ and $T_{i,2}$ (in this order):

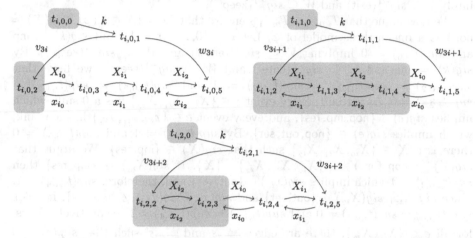

In the following, we argue that H, F_0, F_1 and $G_0, \ldots, G_{m-2}$ collaborate like this: If (sup, sig) is a τ-region solving α then either $sig(k) = \mathsf{inp}$, $V \subseteq sig^{-1}(\mathsf{enter})$ and $W \subseteq sig^{-1}(\mathsf{keep}^-)$ or $sig(k) = \mathsf{out}$ and $V \subseteq sig^{-1}(\mathsf{exit})$ and $W \subseteq sig^{-1}(\mathsf{keep}^+)$. Moreover, we prove that this implies by the functionality of $T_{0,0}, \ldots, T_{m-1,2}$ that $M = \{X \in V(\varphi) \mid sig(X) \neq \mathsf{nop}\}$ is a one-in-three model of φ.

Let (sup, sig) be a τ-region that solves α. Since the interactions $\mathsf{res}, \mathsf{set}$ and nop are defined on both 0 and 1, this implies $sig(k) \in \{\mathsf{inp}, \mathsf{out}, \mathsf{used}, \mathsf{free}\}$. If $sig(k) = \mathsf{used}$ then $sup(s) = sup(s') = 1$ for every transition $s \xrightarrow{k} s'$. Hence, we have $sup(f_{0,3}) = sup(f_{1,1}) = sup(h_{0,4}) = 1$. By definition of $\mathsf{inp}, \mathsf{res}$ we have that $\xrightarrow{e} s$ and $sig(e) \in \{\mathsf{inp}, \mathsf{res}\}$ implies $sup(s) = 0$. Consequently, by $\xrightarrow{z_0} f_{0,3}$ and $\xrightarrow{q_0} f_{1,1}$ we get $sig(z_0), sig(q_0) \in \mathsf{keep}^+$ and thus $sup(h_{0,4}) = sup(h_{0,5}) = sup(h_{0,6}) = 1$ which contradicts $\neg sup(h_{0,6}) \xrightarrow{sig(k)}$. Hence, $sig(k) \neq \mathsf{used}$. Similarly, $sig(k) = \mathsf{free}$ implies $sup(h_{0,6}) = 0$, which is a contradiction. Thus, we have that $sig(k) = \mathsf{inp}$ and $sup(h_{0,6}) = 0$ or $sig(k) = \mathsf{out}$ and $sup(h_{0,6}) = 1$.

As a next step, we show that $sig(k) = \mathsf{inp}$ and $sup(h_{0,6}) = 0$ together imply $sig(v_0) \in \mathsf{enter}$ and $sig(z_0) \in \mathsf{keep}^-$. By $sig(k) = \mathsf{inp}$ and $\xrightarrow{k} h_{0,1}$ and $h_{0,3} \xrightarrow{k}$ we get $sup(h_{0,1}) = 0$ and $sup(h_{0,3}) = 1$. Moreover, by $\xrightarrow{z_0} h_{0,6}$ and $sup(h_{0,6}) = 0$ we obtain $sig(z_0) \in \mathsf{keep}^-$, which by $sup(h_{0,1}) = 0$ implies $sup(h_{0,2}) = 0$.

Finally, $sup(h_{0,2}) = 0$ and $sup(h_{0,3}) = 1$ imply $sig(v_0) \in$ enter. Notice that this reasoning purely bases on $sig(k) =$ inp and $sup(h_{0,6}) = 0$. Moreover, A_φ^τ uses for every $j \in \{0, \ldots, 6m - 2\}$ the TS G_j to ensure $sup(h_{0,6}) = sup(h_{1,6}) = \cdots = sup(h_{6m-1,6})$. This transfers $z_0 \in$ keep$^-$ and $v_0 \in$ enter to $V \subseteq$ enter and $W \subseteq$ keep$^-$. In particular, by $sig(k) =$ inp we have $sup(g_{i,0}) = sup(g_{i,1}) = 1$ and $sup(g_{i,2}) = sup(g_{i,3}) = 0$, that is, $sig(c_i) =$ nop. Hence, if $sig(k) =$ inp and $sup(h_{0,6}) = 0$ then $sup(h_{i,6}) = 0$ for all $i \in \{0, \ldots, 6m - 1\}$. Perfectly similar to the discussion for z_0 and v_0 we obtain that $V \subseteq sig^{-1}($enter$)$ and $W \subseteq sig^{-1}($keep$^-)$, respectively. Similarly, $sig(k) =$ out and $sup(h_{0,6}) = 1$ imply $V \subseteq sig^{-1}($exit$)$ and $W \subseteq sig^{-1}($keep$^+)$.

We now argue that $T_{i,0}, \ldots, T_{m-1,2}$ ensure that $M = \{X \in V(\varphi) \mid sig(X) \neq$ nop$\}$ is a one-in-three model of φ. Let $i \in \{0, \ldots, m - 1\}$ and $sig(k) =$ inp and $sup(h_{0,6}) = 0$ implying $V \subseteq sig^{-1}($enter$)$ and $W \subseteq sig^{-1}($keep$^-)$. By $sig(k) =$ inp and $V \subseteq sig^{-1}($enter$)$ and $W \subseteq sig^{-1}($keep$^-)$ we have that $sup(t_{i,0,2}) = sup(t_{i,1,2}) = sup(t_{i,2,2}) = 1$ and $sup(t_{i,0,5}) = sup(t_{i,1,5}) = sup(t_{i,2,5}) = 0$. As a result, every event $e \in \{X_{i_0}, X_{i_1}, X_{i_2}\}$ has a 0-sink, which implies $sig(e) \in \{$nop, inp, res$\}$, and every event $e \in \{x_{i_0}, x_{i_1}, x_{i_2}\}$ has a 1-sink, which implies $sig(e) \in \{$nop, out, set$\}$. By $sup(t_{i,0,2}) = 1$ and $sup(t_{i,0,5}) = 0$ there is a $X \in \{X_{i_0}, X_{i_1}, X_{i_2}\}$ such that $sig(X) \in \{$inp, res$\}$. We argue that $sig(Y) =$ nop for $Y \in \{X_{i_0}, X_{i_1}, X_{i_2}\} \setminus \{X\}$. If $sig(X_{i_0}) \in \{$inp, res$\}$ then $sup(t_{i,0,3}) = 0$ which implies $sig(x_{i_0}) \in \{$out, set$\}$ and, therefore, $sup(t_{i,1,4}) = 1$. Since $sig(X_{i_1}), sig(X_{i_2}) \notin \{out, set\}$ and $sig(x_{i_1}), sig(x_{i_2}) \notin \{inp, res\}$, it holds $sup(t_{i,0,3}) = sup(t_{i,0,4}) = 0$ and $sup(t_{i,1,3}) = sup(t_{i,1,4}) = 1$, respectively. Thus, for all $e \in \{X_{i_1}, X_{i_2}\}$, there are edges $\xrightarrow{e} s$ and $\xrightarrow{e} s'$ such that $sup(s) = 0$ and $sup(s') = 1$. This implies $sig(e) =$ nop. Similarly, if $sig(X_{i_1}) \in \{$inp, res$\}$, then $sig(X_{i_0}) = sig(X_{i_2}) =$ nop, and if $sig(X_{i_2}) \in \{$inp, res$\}$, then $sig(X_{i_0}) = sig(X_{i_1}) =$ nop. Hence, every clause ζ_i has exactly one variable event with a signature different from nop. This makes $M = \{X \in V(\varphi) \mid sig(X) \neq$ nop$\}$ a one-in-three model of φ. Similarly, if $sig(k) =$ out and $sup(h_{0,6}) = 1$, then M is also a one-in-three model of φ.

To join the gadgets and finally build A_φ^τ, we use the states $\bot = \{\bot_0, \ldots, \bot_{9m+1}\}$ and the events $\oplus = \{\oplus_0, \ldots, \oplus_{9m+1}\}$ and $\ominus = \{\ominus_1, \ldots, \ominus_{9m+1}\}$. The states of $\bot$ are connected by $\bot_j \xrightarrow{\ominus_{j+1}} \bot_{j+1}$ for $j \in \{0, \ldots, 9m + 1\}$. Let $x = 6m + 2$. For all $i \in \{0, \ldots, 6m - 2\}$, for all $j \in \{0, \ldots, m - 1\}$ and for all $\ell \in \{0, 1, 2\}$ we add the following edges that connect the gadgets $H_0, F_0, F_1, G_0, \ldots, G_{6m-2}$ and $T_{0,0}, T_{0,1}, T_{0,2}$ up to $T_{m-1,0}, T_{m-1,1}, T_{m-1,2}$ to A_φ^τ:

$$\bot_0 \xrightarrow{\oplus_0} h_{0,0} \quad \bot_1 \xrightarrow{\oplus_1} f_{0,0} \quad \bot_2 \xrightarrow{\oplus_2} f_{1,0} \quad \bot_{i+3} \xrightarrow{\oplus_{i+3}} g_{i+3,0} \quad \bot_{x+j+3\ell} \xrightarrow{\oplus_{x+j+3\ell}} t_{j,\ell,0}$$

If M is a one-in-three model of φ then there is a τ-region (sup, sig) of A_φ^τ that solves α. The red colored area of the figures introducing the gadgets indicates already a positive support of some states. In particular, if $s \in \{h_{j,0}, h_{j,3} \mid j \in \{0, \ldots, 6m - 1\}\}$ or $s \in \{f_{0,0}, f_{0,2}, f_{0,3}, f_{1,0}, f_{1,1}\}$ $s \in \{g_{j,0}, g_{j,1} \mid j \in \{0, \ldots, 6m - 2\}\}$ then $sup(s) = 1$. The support values of the states of $T_{i,0}, \ldots, T_{i,2}$, where

$i \in \{0, \ldots, m-1\}$, are defined in accordance to which of the events $X_{i_0}, X_{i_1}, X_{i_2}$ belongs to M. The red colored area above sketches the situation where $X_{i_0} \in M$. Moreover, for all $s \in \bot$, we define $sup(s) = 0$. Let $e \in E(A^\tau_\varphi) \setminus \oplus$. We define $sig(e) = \text{inp}$ if $e \in \{k\} \cup M$. For all $i \in \{0, \ldots, m-1\}$ and clauses $\{X_{i_0}, X_{i_1}, X_{i_2}\}$ and all $j \in \{0, 1, 2\}$ we set $sig(e) = \text{set}$ if $e = n$ or $e \in \{v_j, p_j \mid j \in \{0, \ldots, 3m-1\}\}$ or $e = x_{i_j}$ and $X_{i_j} \in M$. Otherwise, we define $sig(e) = \text{nop}$. Finally, for all events $e \in \oplus$ and edges $s \xrightarrow{e} s'$ of A we define $sig(e) = \text{set}$ if $sup(s') = 1$ and, otherwise, $sig(e) = \text{nop}$.

Joining of 1-Bounded Gadgets. In the following, we consider types τ where τ-synthesis from 1-bounded inputs is NP-complete. All gadgets $A_0, \ldots, A_n$ of the reductions are directed paths $A_i = s^i_0 \xrightarrow{e_1} \ldots, \xrightarrow{e_n} s^i_n$ on pairwise distinct states $s^i_0, \ldots, s^i_n$. For all types, the joining is the concatenation

$$A^\tau_\varphi = \quad A_0 \xrightarrow{\ominus_1} \bot_1 \xrightarrow{\oplus_1} A_1 \xrightarrow{\ominus_2} \bot_2 \xrightarrow{\oplus_2} \quad \cdots \quad \xrightarrow{\ominus_n} \bot_n \xrightarrow{\oplus_n} A_n$$

with fresh states $\bot_1, \ldots, \bot_n$ and events $\ominus_1, \cdots \ominus_n, \oplus_1, \cdots \oplus_n$.

Theorem 4. *For any fixed $g \geq 1$, deciding if a g-bounded TS A is τ-solvable is NP-complete if $\tau = \{\text{nop}, \text{inp}, \text{out}, \text{set}\} \cup \omega$ or $\tau = \{\text{nop}, \text{inp}, \text{out}, \text{res}\} \cup \omega$ and $\omega \subseteq \{\text{used}, \text{free}\}$.*

Proof. Our construction proves the claim for $\tau = \{\text{nop}, \text{inp}, \text{set}, \text{out}\} \cup \omega$ with $\omega \subseteq \{\text{used}, \text{free}\}$. By Lemma 2, this proves the claim also for the other types.

The TS A^τ_φ has the following gadgets H_0, H_1, H_2 and H_3 (in this order):

$$h_{0,0} \xrightarrow{k_0} h_{0,1} \xrightarrow{z_0} h_{0,2} \xrightarrow{o} h_{0,3} \xrightarrow{k_1} h_{0,4} \xrightarrow{z_1} h_{0,5} \xrightarrow{z_0} h_{0,6} \xrightarrow{o} h_{0,7} \xrightarrow{k_0} h_{0,8}$$

$$h_{1,0} \xrightarrow{z_0} h_{1,1} \xrightarrow{k_0} h_{1,2} \qquad h_{2,0} \xrightarrow{z_1} h_{2,1} \xrightarrow{k_0} h_{2,2} \qquad h_{3,0} \xrightarrow{k_0} h_{3,1} \xrightarrow{k_1} h_{3,2}$$

If $\text{used} \in \tau$ then A^τ_φ has the following gadget H_4:

$$h_{4,0} \xrightarrow{k_1} h_{4,1} \xrightarrow{z_0} h_{4,2} \xrightarrow{k_1} h_{4,3}$$

For all $i \in \{0, \ldots, m-1\}$, the TS A^τ_φ has for the clause $\zeta_i = \{X_{i_0}, X_{i_1}, X_{i_2}\}$ and the variable $X_i \in V(\varphi)$ the following gadgets T_i and B_i, respectively:

$$t_{i,0} \xrightarrow{k_1} t_{i,1} \xrightarrow{X_{i_0}} t_{i,2} \xrightarrow{X_{i_1}} t_{i,3} \xrightarrow{X_{i_2}} t_{i,4} \xrightarrow{k_0} t_{i,5} \qquad b_{i,0} \xrightarrow{X_i} b_{i,1} \xrightarrow{k_0} b_{i,2}$$

The gadget H_0 provides the atom $\alpha = (k_0, h_{0,6})$. Moreover, the gadgets $H_0, \ldots, H_4$ ensure that if (sup, sig) is a τ-region solving α then $sig(k_0) = \text{out}$ and $sig(k_1) \in \{\text{out}, \text{set}\}$. In particular, H_4 prevents the solvability of α by used. As a result, such a region implies $sup(t_{i,1}) = 1$, $sup(t_{i,4}) = 0$ and $sup(b_{i,1}) = 0$ for all $i \in \{0, \ldots, m-1\}$. On the one hand, by $sup(b_{i,1}) = 0$ for all $i \in \{0, \ldots, m-1\}$ we have $sig(X) \notin \{\text{out}, \text{set}\}$ for all $X \in V(\varphi)$. On the other hand, by $sup(t_{i,1}) = 1$ and $sup(t_{i,4}) = 0$ there is an event $X \in \{X_{i_0}, X_{i_1}, X_{i_2}\}$

such that $sig(X) =$ inp. Since no variable event has an incoming signature we obtain immediately $sig(Y) \neq$ inp for $Y \in \{X_{i_0}, X_{i_1}, X_{i_2}\} \setminus \{X\}$. Thus, $M = \{X \in V(\varphi) \mid sig(X) = \text{inp}\}$ is a one-in-three model of φ.

We argue that $H_0, \ldots, H_4$ behave as announced. Let (sup, sig) be a region that solves $(k_0, h_{0,6})$. If $sig(k_0) =$ inp then $sup(h_{0,6}) = 0$ and $sig(h_{0,7}) = 1$, implying $sig(o) \in \{\text{out}, \text{set}\}$ and $sup(h_{0,3}) = 1$. Thus, there is an event $e \in \{k_1, z_0, z_1\}$ with $sig(e) =$ inp. By $sig(k_0) =$ inp we have $sup(h_{1,1}) = sup(h_{2,1}) = 1$ and $sup(h_{3,1}) = 0$ implying $sig(e) \neq$ inp for all $e \in \{k_1, z_0, z_1\}$, a contradiction.

If $sig(k_0) =$ free then $sup(h_{0,6}) = 1$ and $sup(h_{0,1}) = sup(h_{0,7}) = sup(h_{1,1}) = 0$ which implies $sig(o) =$ inp and $sup(h_{0,2}) = 1$. By $sup(h_{0,1}) = 0$ and $sup(h_{0,2}) = 1$ we have $sig(z_0) \in \{\text{out}, \text{set}\}$ which by $sup(h_{1,1}) = 0$ is a contradiction.

If $sig(k_0) =$ used then $sup(h_{0,6}) = 0$ and $sup(h_{0,1}) = sup(h_{0,7}) = sup(h_{1,1}) = sup(h_{2,1}) = 1$. This implies $sig(o) \in \{\text{out}, \text{set}\}$ and $sup(h_{0,3}) = 1$. Thus, by $sup(h_{0,6}) = 0$ there is an event $e \in \{k_1, z_0, z_1\}$ with $sig(e) =$ inp. By $sup(h_{1,1}) = sup(h_{2,1}) = 1$, we have $e \notin \{z_0, z_1\}$. If $sig(k_1) =$ inp then $sup(h_{4,1}) = 0$ and $sup(h_{4,2}) = 1$, implying $sig(z_0) \in \{\text{out}, \text{set}\}$ and $sup(h_{0,6}) = 1$. This is a contradiction. Altogether, this proves $sig(k_0) \notin \{\text{inp}, \text{used}, \text{free}\}$.

Consequently, we obtain $sig(k_0) =$ out and $sup(h_{0,6}) = 1$ which implies $sig(o) =$ inp and $sup(h_{0,3}) = 0$. By $sup(h_{0,6}) = 1$, this implies that there is an event $e \in \{k_1, z_0, z_1\}$ with $sig(e) \in \{\text{out}, \text{set}\}$. Again by $sig(k_0) =$ out, we have $sup(h_{1,1}) = sup(h_{2,1}) = 0$, which implies $e = k_1$. The signatures $sig(k_0) =$ out and $sig(k_1) \in \{\text{out}, \text{set}\}$ and the construction of $T_0, \ldots, T_{m-1}$ and $B_0, \ldots, B_{m-1}$ ensure that $M = \{X \in V(\varphi) \mid sig(X) = \text{inp}\}$ is a one-in-three model of φ: By $sig(k_0) =$ out and $sig(k_1) \in \{\text{out}, \text{set}\}$ we have $sup(t_{i,1}) = 1$ and $sup(t_{i,4}) = sup(b_{i,1}) = 0$ for all $i \in \{0, \ldots, m-1\}$. By $sup(t_{i,1}) = 1$ and $sup(t_{i,4}) = 0$, there is an event $X \in \zeta_i$ such that $sig(X) =$ inp. Moreover, by $sup(b_{i,1}) = 0$, we get $sig(X_i) \notin$ enter for all $i \in \{0, \ldots, m-1\}$. Thus, X is unambiguous and thus M is a searched model.

Conversely, if M is a one-in-three model of φ then there is a τ-region (sup, sig) that solves α. The red colored area above sketches states with a positive support. Which states of T_i, besides of $t_{i,0}, t_{i,1}$ and $t_{i,5}$, get a positive support depends for all $i \in \{0, \ldots, m-1\}$ on which of $X_{i_0}, X_{i_1}, X_{i_2}$ belongs to M. The red colored area above sketches the case $X_{i_0} \in M$. Moreover, we define $sup(s) = 1$ if $s = b_{i,0}$ and $X_i \in M$ or if $s \in \bot$. The signature is defined as follows: $sig(k_1) =$ set; for all $e \in E(A_\varphi^\tau) \setminus \{k_1\}$ and all $s \xrightarrow{e} s' \in A_\varphi^\tau$, if $sup(s') > sup(s)$, then $sig(e) =$ out; if $sup(s) > sup(s')$, then $sig(e) =$ inp; else $sig(e) =$ nop. $\square$

Theorem 5. *For any $g \geq 1$, deciding if a g-bounded TS A is τ-solvable is NP-complete if $\tau = \{\text{nop}, \text{inp}, \text{set}, \text{free}\}$ or $\tau = \{\text{nop}, \text{inp}, \text{set}, \text{used}, \text{free}\}$ or $\tau = \{\text{nop}, \text{out}, \text{res}, \text{used}\}$ or $\tau = \{\text{nop}, \text{out}, \text{res}, \text{used}, \text{free}\}$.*

Proof. Our reduction proves the claim for $\tau = \{\text{nop}, \text{inp}, \text{set}, \text{free}\}$ and $\tau = \{\text{nop}, \text{inp}, \text{set}, \text{used}, \text{free}\}$ and thus by Lemma 2, for the other types, too.

The TS A_φ^τ has the following gadgets H_0 and H_1 providing the atom $(k_0, h_{0,3})$:

$$h_{0,0} \xrightarrow{k_0} h_{0,1} \xrightarrow{k_1} h_{0,2} \xrightarrow{z_0} h_{0,3} \xrightarrow{k_1} h_{0,4} \xrightarrow{z_1} h_{0,5} \xrightarrow{k_0} h_{0,6}$$

$$h_{1,0} \xrightarrow{k_0} h_{1,1} \xrightarrow{z_0} h_{1,2} \xrightarrow{k_0} h_{1,3}$$

For all $i \in \{0, \ldots, m-1\}$, the A_φ^τ for the clause $\zeta_i = \{X_{i_0}, X_{i_1}, X_{i_2}\}$ and the variable $X_i \in V(\varphi)$ the gadgets T_i and B_i as previously defined for Theorem 4. The gadgets H_0 and H_1 ensure that a τ-region (sup, sig) solving $(k_0, h_{0,3})$ satisfies $sig(k_0) =$ free and $sig(k_1) =$ set. This implies $sup(t_{i,1}) = 1$ and $sup(t_{i,4}) = sup(b_{i,2}) = 0$ for all $i \in \{0, \ldots, m-1\}$. By $sup(t_{i,1}) = 1$ and $sup(t_{i,4}) = 0$, there is an event $X \in \zeta_i$ such that $sig(X) =$ inp and, by $sup(b_{i,2}) = 0$ for all $i \in \{0, \ldots, m-1\}$, we have $sig(X) \neq$ set for all $X \in V(\varphi)$. Thus, the event $X \in \zeta_i$ is unique and $M = \{X \in V(\varphi) \mid sig(X) = \text{inp}\}$ is a one-in-three model.

We briefly argue that H_0 and H_1 perform as announced: Let (sup, sig) be a τ-region that solves α. If $sig(k_0) =$ inp then $sup(h_{1,1}) = 0$ and $sup(h_{1,2}) = 1$ which implies $sig(z_0) =$ set and thus $sup(h_{0,3}) = 1$, a contradiction. Hence, $sig(k_0) \neq$ inp. If $sig(k_0) =$ used then $sup(h_{0,1}) = sup(h_{1,2}) = 1$ and $sup(h_{0,3}) = 0$. Consequently, $sig(z_0) =$ inp or $sig(k_1) =$ inp but this contradicts $sup(h_{1,2}) = 1$ and $sup(h_{0,3}) = 0$. Thus, $sig(k_0) \neq$ used. Thus, we have $sig(k_0) =$ free and $sup(h_{0,3}) = 1$, which implies that one of k_1, z_0 has a set-signature. By $sig(k_0) =$ free, we get $sup(h_{1,3}) = 0$ and thus $sig(k_1) =$ set.

If M is a one-in-three model of φ then we can define an α solving region similar to the one of Theorem 4, where we replace $sig(k_0) =$ inp by $sig(k_0) =$ free.

Theorem 6. *For any fixed $g \geq 1$, deciding if a g-bounded TS A is τ-solvable is NP-complete if $\tau = \{nop, inp, res, swap\} \cup \omega$ or $\tau = \{nop, out, set, swap\} \cup \omega$ and $\omega \subseteq \{used, free\}$.*

Proof. The TS A_φ^τ has the following gadgets H_0, H_1, H_2 and H_3:

$$h_{0,0} \xrightarrow{k} h_{0,1} \xrightarrow{y_0} h_{0,2} \xrightarrow{v} h_{0,3} \xrightarrow{k} h_{0,4} \qquad h_{1,0} \xrightarrow{k} h_{1,1} \xrightarrow{y_1} h_{1,2} \xrightarrow{y_0} h_{1,3} \xrightarrow{k} h_{1,4}$$

$$h_{2,0} \xrightarrow{k} h_{2,1} \xrightarrow{y_0} h_{2,2} \xrightarrow{y_1} h_{2,3} \xrightarrow{y_0} h_{2,4} \xrightarrow{k} h_{2,5} \qquad h_{3,0} \xrightarrow{y_1} h_{3,1} \xrightarrow{y_0} h_{3,2} \xrightarrow{v} h_{3,3} \xrightarrow{k} h_{3,4}$$

The gadgets $H_0, \ldots, H_3$ provide the atom $\alpha = (k, h_{0,2})$ and ensure that a τ-region (sup, sig) solving α satisfies $sig(k) =$ inp and $sup(h_{0,2}) = 0$. The TS A_φ^τ has the following gadgets F_0, F_1 and for all $j \in \{0, \ldots, 10\}$ the gadget G_j:

$$f_{0,0} \xrightarrow{k} f_{0,1} \xrightarrow{z_0} f_{0,2} \xrightarrow{v} f_{0,3} \xrightarrow{k} f_{0,4} \qquad f_{1,0} \xrightarrow{k} f_{1,1} \xrightarrow{z_1} f_{1,2} \xrightarrow{v} f_{1,3} \xrightarrow{k} f_{1,4}$$

$$g_{j,0} \xrightarrow{k} g_{j,1} \xrightarrow{z_0} g_{j,2} \xrightarrow{u_j} g_{j,3} \xrightarrow{z_1} g_{j,4} \xrightarrow{k} g_{j,5}$$

For all $j \in \{0, \ldots, 10\}$, the gadgets F_0, F_1, G_j ensure $sig(u_j) =$ swap for any τ-region (sup, sig) solving α.

For all $i \in \{0, \ldots, m-1\}$, the TS A_φ^τ has for the clause $\zeta_i = \{X_{i_0}, X_{i_1}, X_{i_2}\}$ some gadgets $T_{i,0}, \ldots, T_{i,6}$ and B_i.   The purpose of these gadgets is to make the one-and-three satisfiability of φ and the solvability of α the same. In particular, the TS $T_{i,0}$ is defined by:

$$t_{i,0,0} \xrightarrow{k} t_{i,0,1} \xrightarrow{u_0} t_{i,0,2} \xrightarrow{X_{i_0}} t_{i,0,3} \xrightarrow{u_1} t_{i,0,4} \xrightarrow{X_{i_1}} t_{i,0,5} \xrightarrow{u_2} t_{i,0,6} \xrightarrow{X_{i_2}} t_{i,0,7} \xrightarrow{u_3} t_{i,0,8} \xrightarrow{k} t_{i,0,9}$$

The gadgets $T_{i,1}, T_{i,2}$ and $T_{i,3}$ are defined (in this order) as follows:

$$t_{i,1,0} \xrightarrow{k} t_{i,1,1} \xrightarrow{u_4} t_{i,1,2} \xrightarrow{u_5} t_{i,1,3} \xrightarrow{X_{i_0}} t_{i,1,4} \xrightarrow{w_{3i}} t_{i,1,5} \xrightarrow{X_{i_1}} t_{i,1,6} \xrightarrow{u_6} t_{i,1,7} \xrightarrow{k} t_{i,1,8}$$

$$t_{i,2,0} \xrightarrow{k} t_{i,2,1} \xrightarrow{u_4} t_{i,2,2} \xrightarrow{u_5} t_{i,2,3} \xrightarrow{X_{i_2}} t_{i,2,4} \xrightarrow{w_{3i+1}} t_{i,2,5} \xrightarrow{X_{i_0}} t_{i,2,6} \xrightarrow{u_6} t_{i,2,7} \xrightarrow{k} t_{i,2,8}$$

$$t_{i,3,0} \xrightarrow{k} t_{i,3,1} \xrightarrow{u_4} t_{i,3,2} \xrightarrow{u_5} t_{i,3,3} \xrightarrow{X_{i_1}} t_{i,3,4} \xrightarrow{w_{3i+2}} t_{i,3,5} \xrightarrow{X_{i_2}} t_{i,3,6} \xrightarrow{u_6} t_{i,3,7} \xrightarrow{k} t_{i,3,8}$$

Moreover, the gadgets $T_{i,4}, T_{i,5}$ and $T_{i,6}$ are defined like this:

$$t_{i,4,0} \xrightarrow{k} t_{i,4,1} \xrightarrow{u_7} t_{i,4,2} \xrightarrow{w_{3i}} t_{i,4,3} \xrightarrow{u_8} t_{i,4,4} \xrightarrow{k} t_{i,4,5}$$

$$t_{i,5,0} \xrightarrow{k} t_{i,5,1} \xrightarrow{u_7} t_{i,5,2} \xrightarrow{w_{3i+1}} t_{i,5,3} \xrightarrow{u_8} t_{i,5,4} \xrightarrow{k} t_{i,5,5}$$

$$t_{i,6,0} \xrightarrow{k} t_{i,6,1} \xrightarrow{u_7} t_{i,6,2} \xrightarrow{w_{3i+2}} t_{i,6,3} \xrightarrow{u_8} t_{i,6,4} \xrightarrow{k} t_{i,6,5}$$

Finally, the gadget B_i is defined as follows:

$$b_{i,0} \xrightarrow{X_i} b_{i,1} \xrightarrow{u_9} b_{i,2} \xrightarrow{u_{10}} b_{i,3} \xrightarrow{k} b_{i,4}$$

Let (sup, sig) be a τ-region solving α. We first argue that the gadgets $H_0, \ldots, H_3$ and F_0, F_1 and $G_0, \ldots, G_{10}$ ensure that a τ-region (sup, sig) solving α satisfies $sig(k) = \mathsf{inp}$, $sup(h_{0,2}) = 0$ and $sig(u_0) = \cdots = sig(u_{10}) = \mathsf{swap}$.

If $sig(k) = \mathsf{free}$ and $sup(h_{0,2}) = 1$ then $s \xrightarrow{k} s'$ implies $sup(s) = sup(s') = 0$. Especially, by $sup(h_{0,1}) = 0$ and $sup(h_{0,2}) = 1$ we have $sig(y_0) = \mathsf{swap}$. Moreover, by $sup(h_{2,1}) = sup(h_{2,4}) = 0$ and $sig(y_0) = \mathsf{swap}$ we have that $sup(h_{2,2}) = sup(h_{2,3}) = 1$. This implies $sig(y_1) \in \{\mathsf{nop}, \mathsf{used}\}$. By $sup(h_{1,1}) = 0$ and $h_{1,1} \xrightarrow{y_1}$ this implies $sig(y_1) = \mathsf{nop}$ and thus $sup(h_{1,2}) = 0$. Furthermore, by $sup(h_{1,2}) = sup(h_{1,3}) = 0$ and $h_{1,2} \xrightarrow{y_0} h_{1,3}$ this implies $sig(y_0) \neq \mathsf{swap}$, a contradiction. Thus, we have $sig(k) \neq \mathsf{free}$.

If $sig(k) = \mathsf{used}$ and $sup(h_{0,2}) = 0$ then $s \xrightarrow{k} s'$ implies $sup(s) = sup(s') = 1$. Thus, we get $sup(h_{0,1}) = sup(h_{0,3}) = sup(h_{1,3}) = 1$ which with $sup(h_{0,2}) = 0$ implies $sig(y_0) = sig(v) = \mathsf{swap}$. Moreover, $sup(h_{1,3}) = 1$ and $sig(y_0) = \mathsf{swap}$ imply $sup(h_{1,2}) = 0$. By $sup(h_{1,1}) = 1$, this implies $sig(y_1) \in \{\mathsf{inp}, \mathsf{res}\}$. Finally, $sup(h_{3,3}) = 1$ and $sig(v) = sig(y_0) = \mathsf{swap}$ imply $sup(h_{3,1}) = 1$. This contradicts

$sig(y_1) \in \{\mathsf{inp}, \mathsf{res}\}$. Thus, $sig(k) \neq \mathsf{used}$. Altogether, this shows that $sig(k) = \mathsf{inp}$ and $sup(h_{0,2}) = 0$, which implies $sig(v) = \mathsf{swap}$.

By $sig(k) = \mathsf{inp}$ we have $sup(f_{0,1}) = sup(f_{1,1}) = sup(g_{j,1}) = 0$ and $sup(f_{0,3}) = sup(f_{1,3}) = sup(g_{j,4}) = 1$. By $sig(v) = \mathsf{swap}$, this implies $sup(f_{0,2}) = sup(f_{1,2}) = 0$ and thus $sig(z_0), sig(z_1) \in \{\mathsf{nop}, \mathsf{res}, \mathsf{free}\}$. Moreover, $sup(g_{j,1}) = 0$, $sup(g_{j,4}) = 1$ and $sig(z_0), sig(z_1) \in \{\mathsf{nop}, \mathsf{res}, \mathsf{free}\}$ imply $sup(g_{j,2}) = 0$ and $sup(g_{j,3}) = 1$ and thus $sig(u_j) = \mathsf{swap}$.

Let $i \in \{0, \ldots, m-1\}$. We now show that $T_{i,0}, \ldots, T_{i,6}$ and B_i collaborate as announced. By $sig(k) = \mathsf{inp}$ and $sig(u_9) = sig(u_{10}) = \mathsf{swap}$, we have $sup(b_{i,1}) = 1$ for all $i \in \{0, \ldots, m-1\}$. Since $\xrightarrow{X_i} b_{i,1}$ for all $i \in \{0, \ldots, m-1\}$, the gadget B_i ensures for all $X \in V(\varphi)$ that $s \xrightarrow{X} s'$ and $sup(s) \neq sup(s')$ imply $sig(X) = \mathsf{swap}$. The gadget $T_{i,0}$ works like this: By $sig(k) = \mathsf{inp}$ we get that $sup(t_{i,0,1}) = 0$ and $sup(t_{i,0,8}) = 1$. Consequently, the image $sup(t_{i,0,1}) \xrightarrow{sig(X_{i_0})} \ldots \xrightarrow{sig(u_3)} sup(t_{i,0,8})$ of the path $t_{i,0,1} \xrightarrow{X_{i_0}} \ldots \xrightarrow{u_3} t_{i,0,8}$ performs an odd number of state changes between 0 to 1 in τ. Since $sig(u_0) = \cdots = sig(u_3) = \mathsf{swap}$, the events $u_0, \ldots, u_3$ perform an even number of state changes. Thus, either all of $X_{i_0}, X_{i_1}, X_{i_2}$ are mapped to swap or exactly one of them. The construction of $T_{i,1}, \ldots, T_{i,6}$ guarantees that there is exactly one variable event mapped to swap.

In particular, the gadgets $T_{i,4}, T_{i,5}$ and $T_{i,6}$ ensure that if $e \in \{w_{3i}, w_{3i+1}, w_{3i+2}\}$ then $sig(e) \notin \{\mathsf{nop}, \mathsf{used}\}$. We argue for w_{3i}: By $sig(k) = \mathsf{inp}$ we get $sup(t_{i,4,1}) = 0$ and $sup(t_{i,4,4}) = 1$ which, by $sig(u_7) = sig(u_8) = \mathsf{swap}$, implies $sup(t_{i,4,2}) = 1$ and $sup(t_{i,4,3}) = 0$. Clearly, this implies $sig(w_{3i}) \notin \{\mathsf{nop}, \mathsf{used}\}$. Similarly, we obtain that $sig(w_{3i+1}) \notin \{\mathsf{nop}, \mathsf{used}\}$ and $sig(w_{3i+2}) \notin \{\mathsf{nop}, \mathsf{used}\}$.

Finally, the gadgets $T_{i,1}, T_{i,2}$ and $T_{i,3}$ ensure that no two variable events of the same clause can have a swap signature: By $sig(k) = \mathsf{inp}$ we get that $sup(t_{i,1,1}) = 0$ and $sup(t_{i,1,7}) = 1$ which with $sig(u_4) = sig(u_5) = sig(u_6) = \mathsf{swap}$ implies $sup(t_{i,1,3}) = 0$ and $sup(t_{i,1,6}) = 0$. Thus, if $sig(X_{i_0}) = sig(X_{i_1}) = \mathsf{swap}$ then $sup(t_{i,1,4}) = sup(t_{i,1,5}) = 1$ which implies $sig(w_{3i}) \in \{\mathsf{nop}, \mathsf{used}\}$, a contradiction. Similarly, one uses $T_{i,2}$ and $T_{i,3}$ to show that neither X_{i_0} and X_{i_2} nor X_{i_1} and X_{i_2} can simultaneously be mapped to swap. As i was arbitrary, there is exactly one variable per clause that is mapped to swap. Thus, $M = \{X \in V(\varphi) \mid sig(X) = \mathsf{swap}\}$ is a one-in-three model of φ.

Conversely, a one-in-three model M of φ allows a τ-region (sup, sig) that solves α: The red colored area above indicates which states of $H_0, \ldots, H_3, F_0, F_1, G_0, \ldots, G_{10}$ and $T_{0,4}, T_{0,5}, T_{0,6}, \ldots, T_{m-1,4}, T_{m-1,5}, T_{m-1,6}$ have positive support. Moreover, we define $sup(s) = 1$ for all $s \in \bot$. Which states of $T_{i,0}, \ldots, T_{i,3}$, where $i \in \{0, \ldots, m-1\}$, besides of k's sources get a positive support depends on which of $X_{i_0}, X_{i_1}, X_{i_2}$ belongs to M. The red colored area sketches the situation for $X_{i_0} \in M$. It is easy to see that there is for all $e \in E(A_\varphi^\tau)$ a fitting sig-value making (sup, sig) a (solving) τ-region where $sig(k) = \mathsf{inp}$ and $sup(h_{0,2}) = 0$. $\square$

Theorem 7. *For any fixed $g \geq 1$, deciding if a g-bounded TS A is τ-solvable is NP-complete if $\tau = \{nop, inp, set, swap\} \cup \omega$ and $\omega \subseteq \{out, res, used, free\}$ or if $\tau = \{nop, out, res, swap\} \cup \omega$ and $\omega \subseteq \{inp, set, used, free\}$.*

Proof. We present the reduction for the types built by $\tau = \{nop, inp, set, swap\} \cup \omega$ where $\omega \subseteq \{out, res, used, free\}$. Again, the other types are covered by Lemma 2.

The TS A_φ^τ has the following gadgets H_0, H_1, H_2 and H_3:

$$h_{0,0} \xrightarrow{k} h_{0,1} \xrightarrow{v_0} h_{0,2} \qquad\qquad h_{1,0} \xrightarrow{v_0} h_{1,1} \xrightarrow{k} h_{1,2}$$

$$h_{2,0} \xrightarrow{k} h_{2,1} \xrightarrow{v_0} h_{2,2} \xrightarrow{v_1} h_{2,3} \xrightarrow{k} h_{2,4} \qquad h_{3,0} \xrightarrow{k} h_{3,1} \xrightarrow{v_1} h_{3,2} \xrightarrow{v_0} h_{3,3}$$

If $\tau \cap \{used, free\} \neq \emptyset$ then A_φ^τ has also the following gadgets $H_4, \ldots, H_{12}$:

$$h_{4,0} \xrightarrow{k} h_{4,1} \xrightarrow{x} h_{4,2} \xrightarrow{v_0} h_{4,3} \xrightarrow{k} h_{4,4} \qquad h_{5,0} \xrightarrow{k} h_{5,1} \xrightarrow{v_0} h_{5,2} \xrightarrow{x} h_{5,3} \xrightarrow{k} h_{5,4}$$

$$h_{6,0} \xrightarrow{k} h_{6,1} \xrightarrow{x} h_{6,2} \xrightarrow{y_0} h_{6,3} \xrightarrow{k} h_{6,4} \qquad h_{7,0} \xrightarrow{k} h_{7,1} \xrightarrow{y_0} h_{7,2} \xrightarrow{x} h_{7,3} \xrightarrow{k} h_{7,4}$$

$$h_{8,0} \xrightarrow{k} h_{8,1} \xrightarrow{x} h_{8,2} \xrightarrow{y_1} h_{8,3} \xrightarrow{k} h_{8,4} \qquad h_{9,0} \xrightarrow{k} h_{9,1} \xrightarrow{y_1} h_{9,2} \xrightarrow{x} h_{9,3} \xrightarrow{k} h_{9,4}$$

$$h_{10,0} \xrightarrow{k} h_{10,1} \xrightarrow{x} h_{10,2} \xrightarrow{y_2} h_{10,3} \xrightarrow{k} h_{10,4} \qquad h_{11,0} \xrightarrow{k} h_{11,1} \xrightarrow{y_2} h_{11,2} \xrightarrow{x} h_{11,3} \xrightarrow{k} h_{11,4}$$

$$h_{12,0} \xrightarrow{k} h_{12,1} \xrightarrow{y_0} h_{12,2} \xrightarrow{y_1} h_{12,3} \xrightarrow{y_2} h_{12,4} \xrightarrow{k} h_{12,5}$$

The gadgets $H_0, \ldots, H_3$ ($H_4, \ldots, H_{12}$, if added) provide $\alpha = (k, h_{3,3})$. They ensure that if (sup, sig) τ-solves α, then $sig(k) \in \{inp, out\}$. The TS A_φ^τ adds the following gadgets F_0, F_1, F_2 and, for all $i \in \{0, \ldots, 13\}$, the gadgets G_i, N_i:

$$f_{0,0} \xrightarrow{k} f_{0,1} \xrightarrow{z_0} f_{0,2} \xrightarrow{v_0} f_{0,3} \xrightarrow{k} f_{0,4} \qquad f_{1,0} \xrightarrow{k} f_{1,1} \xrightarrow{z_1} f_{1,2} \xrightarrow{v_0} f_{1,3} \xrightarrow{k} f_{1,4}$$

$$f_{2,0} \xrightarrow{k} f_{2,1} \xrightarrow{z_0} f_{2,2} \xrightarrow{z_1} f_{2,3} \xrightarrow{z_2} f_{2,4} \xrightarrow{k} f_{2,5}$$

$$g_{i,0} \xrightarrow{k} g_{i,1} \xrightarrow{z_0} g_{i,2} \xrightarrow{u_i} g_{i,3} \xrightarrow{z_1} g_{i,4} \xrightarrow{k} g_{i,5} \qquad n_{i,0} \xrightarrow{k} n_{i,1} \xrightarrow{z_2} n_{i,2} \xrightarrow{u_i} n_{i,3} \xrightarrow{v_0} n_{i,4} \xrightarrow{k} n_{i,5}$$

The gadgets F_0, F_1, F_2 and $G_0, N_0, \ldots, G_{13}, N_{13}$ guarantee that if (sup, sig) solves α then $sig(u_i) = swap$. Similarly to the reduction of Theorem 6, the TS A_φ^τ has for every $i \in \{0, \ldots, m-1\}$ gadgets $T_{i,0}, \ldots, T_{i,6}$ and B_i to make the one-in-three satisfiability of φ and the τ-solvability of α the same. These gadgets and the ones for Theorem 6 have basically the same intention. However, since the current types have different interactions, the peculiarity of these gadgets is slightly different. In particular, A_φ^τ has for each clause $\zeta_i = \{X_{i_0}, X_{i_1}, X_{i_2}\}$ the following gadget $T_{i,0}$:

$$t_{i,0,0} \xrightarrow{k} t_{i,0,1} \xrightarrow{u_0} t_{i,0,2} \xrightarrow{X_{i_0}} t_{i,0,3} \xrightarrow{u_1} t_{i,0,4} \xrightarrow{X_{i_1}} t_{i,0,5} \xrightarrow{u_2} t_{i,0,6} \xrightarrow{X_{i_2}} t_{i,0,7} \xrightarrow{u_3} t_{i,0,8} \xrightarrow{k} t_{i,0,9}$$

Moreover, the gadgets $T_{i,1}, T_{i,2}$ and $T_{i,3}$ are defined as follows:

$$t_{i,1,0} \xrightarrow{k} t_{i,1,1} \xrightarrow{u_4} t_{i,1,2} \xrightarrow{X_{i_0}} t_{i,1,3} \xrightarrow{w_{3i}} t_{i,1,4} \xrightarrow{X_{i_1}} t_{i,1,5} \xrightarrow{u_5} t_{i,1,6} \xrightarrow{u_6} t_{i,1,7} \xrightarrow{k} t_{i,1,8}$$

$$t_{i,2,0} \xrightarrow{k} t_{i,2,1} \xrightarrow{u_4} t_{i,2,2} \xrightarrow{X_{i_2}} t_{i,2,3} \xrightarrow{w_{3i+1}} t_{i,2,4} \xrightarrow{X_{i_0}} t_{i,2,5} \xrightarrow{u_5} t_{i,2,6} \xrightarrow{u_6} t_{i,2,7} \xrightarrow{k} t_{i,2,8}$$

$$t_{i,3,0} \xrightarrow{k} t_{i,3,1} \xrightarrow{u_4} t_{i,3,2} \xrightarrow{X_{i_1}} t_{i,3,3} \xrightarrow{w_{3i+2}} t_{i,3,4} \xrightarrow{X_{i_2}} t_{i,3,5} \xrightarrow{u_5} t_{i,3,6} \xrightarrow{u_6} t_{i,3,7} \xrightarrow{k} t_{i,3,8}$$

Furthermore, the gadgets $T_{i,4}, T_{i,5}$ and $T_{i,6}$ are defined by

$$t_{i,4,0} \xrightarrow{k} t_{i,4,1} \xrightarrow{u_7} t_{i,4,2} \xrightarrow{u_8} t_{i,4,3} \xrightarrow{w_{3i}} t_{i,4,4} \xrightarrow{u_9} t_{i,4,5} \xrightarrow{u_{10}} t_{i,4,6} \xrightarrow{k} t_{i,4,7}$$

$$t_{i,5,0} \xrightarrow{k} t_{i,5,1} \xrightarrow{u_7} t_{i,5,2} \xrightarrow{u_8} t_{i,5,3} \xrightarrow{w_{3i+1}} t_{i,5,4} \xrightarrow{u_9} t_{i,5,5} \xrightarrow{u_{10}} t_{i,5,6} \xrightarrow{k} t_{i,5,7}$$

$$t_{i,6,0} \xrightarrow{k} t_{i,6,1} \xrightarrow{u_7} t_{i,6,2} \xrightarrow{u_8} t_{i,6,3} \xrightarrow{w_{3i+2}} t_{i,6,4} \xrightarrow{u_9} t_{i,6,5} \xrightarrow{u_{10}} t_{i,6,6} \xrightarrow{k} t_{i,6,7}$$

Finally, the TS A_φ^τ has for all $i \in \{0, \ldots, m-1\}$ the following gadget B_i:

$$b_{i,0} \xrightarrow{X_i} b_{i,1} \xrightarrow{u_{11}} b_{i,2} \xrightarrow{k} b_{i,3}$$

We briefly argue for the announced functionality of the gadgets. Let (sup, sig) be a τ-region solving α. If $sig(k) = $ free then $sup(h_{3,3}) = 1$ and $s \xrightarrow{k} s'$ implies $sup(s) = sup(s') = 0$. Since $sup(h_{3,1}) = 0$ and $sup(h_{3,3}) = 1$, there is an event $e \in \{v_0, v_1\}$ such that $sig(e) \in \{\mathsf{out}, \mathsf{set}, \mathsf{swap}\}$. If $sig(v_0) \in \{\mathsf{out}, \mathsf{set}, \mathsf{swap}\}$, then, by $sup(h_{1,1}) = 0$, we get $sig(v_0) = \mathsf{swap}$. Moreover, if $sig(v_1) \in \{\mathsf{out}, \mathsf{set}, \mathsf{swap}\}$, which implies $sig(h_{3,2}) = 1$, then, by $sup(h_{2,3}) = 0$, we get $sig(v_1) = \mathsf{swap}$. By $sig(v_1) = \mathsf{swap}$ and $sup(h_{2,3}) = 0$, we get $sup(h_{2,2}) = 1$. By $sup(h_{1,1})$, this implies $sig(v_0) = \mathsf{swap}$. Thus, in any case we get $sig(v_0) = \mathsf{swap}$. By $sig(v_0) = \mathsf{swap}$ and $sup(h_{4,3}) = sup(h_{5,1}) = 0$ we obtain $sup(h_{4,2}) = sup(h_{5,2}) = 1$ which implies $sig(x) = \mathsf{swap}$. Using this and $sup(s) = sup(s') = 0$ if $s \xrightarrow{k} s'$, we have that $sup(h_{j,2}) = 1$ for all $j \in \{6, \ldots, 11\}$. This implies $sig(y_0) = sig(y_1) = sig(y_2) = \mathsf{swap}$. By $sup(h_{12,1}) = sup(h_{12,4}) = 0$, the image of $h_{12,1} \xrightarrow{y_0} \ldots \xrightarrow{y_2} h_{12,4}$ is a path from 0 to 0 in τ. The number of state changes between 0 and 1 on such a path is even. This contradicts $sig(y_0) = sig(y_1) = sig(y_2) = \mathsf{swap}$. Thus, $sig(k) \neq \mathsf{free}$. The assumption that $sig(k) = \mathsf{used}$ and $sup(h_{3,3}) = 0$ yields a contradiction, too.

We conclude that $sig(k) = \mathsf{inp}$ and $sup(h_{3,3}) = 0$. This implies $sig(v_0) \notin \{\mathsf{out}, \mathsf{set}\}$ and if $s \xrightarrow{k} s' \in A_I^\tau$, then $sup(s) = 1$ and $sup(s') = 0$. Thus, by $sup(h_{2,1}) = 0$ and $sup(h_{2,3}) = 1$ there is an event $e \in \{v_0, v_1\}$ such that $sig(e) \in \{\mathsf{out}, \mathsf{set}, \mathsf{swap}\}$. If $e = v_0$ then $sig(v_0) = \mathsf{swap}$. Moreover, if

$e = v_1$ then $sup(h_{3,2}) = 1$ which with $sup(h_{3,3}) = 0$ and $sup(h_{1,1}) = 1$ implies $sig(v_0) = $ swap. Consequently, any case implies $sig(v_0) = $ swap. This results in $sig(u_j) = $ swap for all $j \in \{0, \ldots, 13\}$ as follows. By $sup(f_{0,3}) = sup(f_{1,3}) = 1$ and $sig(v) = $ swap we obtain $sup(f_{0,2}) = sup(f_{1,2}) = 0$ which with $sup(f_{0,1}) = sup(f_{1,1}) = 0$ implies $sig(z_0), sig(z_1) \in \{$nop, res, free$\}$. Moreover, by $sig(z_0), sig(z_1) \in \{$nop, res, free$\}$ and $sup(f_{2,1}) = 0$ we get $sup(f_{2,3}) = 0$ which with $sup(f_{2,4}) = 1$ implies $sig(z_2) \in \{$out, set, swap$\}$. By $sig(z_0) \in \{$nop, res, free$\}$ and $sup(g_{i,1}) = 0$, we get $sup(g_{i,2}) = 0$. Furthermore, $sig(z_1) \in \{$nop, res, free$\}$ and $sup(g_{i,4}) = 1$ yields $sig(z_1) = $ nop and $sup(g_{i,3}) = 1$. This implies $sig(u_i) \in \{$out, set, swap$\}$. Finally, by $sup(n_{i,1}) = 0$ and $sig(z_2) \in \{$out, set, swap$\}$, we get $sup(n_{i,1}) = 1$ and, by $sup(n_{i,4}) = 1$ and $sig(v_0) = $ swap, we have $sup(n_{i,3}) = 0$. Since $sig(u_i) \in \{$out, set, swap$\}$, this yields $sig(u_i) = $ swap for all $i \in \{0, \ldots, 13\}$. The gadgets $T_{i,0}, \ldots, T_{i,6}$, where $i \in \{0, \ldots, m-1\}$, use $sig(k) = $ inp and $sig(u_i) = $ swap for all $i \in \{0, \ldots, 13\}$ similarly to the ones of Theorem 6 to ensure that $M = \{X \in V(\varphi) \mid sig(X) = $ swap$\}$ is a one-in-three model of φ: By $sup(t_{i,4,6}) = sup(t_{i,5,6}) = sup(t_{i,6,6}) = 1$ and $sig(u_5) = sig(u_6) = $ swap we have $sup(t_{i,4,4}) = sup(t_{i,5,4}) = sup(t_{i,6,4}) = 1$ for all $i \in \{0, \ldots, m-1\}$. Thus, if $X \in V(\varphi)$, $s \xrightarrow{X} s'$ and $sup(s) \neq sup(s')$ then $sig(X) = $ swap. Using this, one argues in a manner quite similar to that of the proof of Theorem 6 that $T_{i,0}, \ldots, T_{i,6}$ collaborate in such a way that there is exactly one variable event $X \in \{X_{i_0}, X_{i_1}, X_{i_2}\}$ such that $sig(X) = $ swap. Thus, M is a corresponding model. Moreover, if $sig(k) = $ out and $sup(h_{3,3}) = 1$ then we obtain again that $sig(u_i) = $ swap for all $i \in \{0, \ldots, 13\}$ which also guarantees that M is a searched model.

Conversely, if M is a one-in-three model of φ then we can define analogously to Theorem 6 a τ-region solving α.   □

Theorem 8 ([12]). *For any fixed $g \geq 1$, deciding if a g-bounded TS A is τ-solvable is NP-complete if $\tau \in \{$nop, inp, out$\} \cup \{$used, free$\}$.*

Proof. The claim follows directly from our result of [12]. There we use 1-bounded cycle free gadgets to prove that synthesis of (pure) b-bounded Petri nets is NP-complete. The joining of [12] yields a 2-bounded TS. However, it is easy to see that the 1-bounded joining of this paper fits, too. The (pure) 1-bounded Petri net type is isomorphic to $\{$nop, inp, out, used$\}$ ($\{$nop, inp, out$\}$). By symmetry, τ-solving ESSP atoms by used is equivalent to solving them by free.   □

4   Polynomial Time Results

Theorem 9. *For any fixed $g < 2$, one can decide in polynomial time if a g-bounded TS A is τ-solvable if $\tau = \{$nop, inp, set$\}$ or $\tau = \{$nop, inp, set, used$\}$ or $\tau = \{$nop, out, res$\}$ or $\tau = \{$nop, out, res, free$\}$ or $\tau = \{$nop, set, res$\} \cup \omega$ with non-empty $\omega \subseteq \{$inp, out, used, free$\}$.*

Proof. If A is τ-solvable then no event e of A occurs twice in a row. Otherwise, the SSP atom (s', s'') of a sequence $s \xrightarrow{e} s' \xrightarrow{e} s''$ is not τ-solvable. Thus, in what follows, we assume that A has no event occurring twice in a row. Moreover, it is easy to see that a 1-bounded TS $A = s_0 \xrightarrow{e_1} \ldots \xrightarrow{e_m} s_m$ is a simple directed path on pairwise distinct states $s_0, \ldots, s_m$ or a directed cycle, that is, all states $s_0, \ldots, s_m$ except s_0 and s_m are pairwise distinct. This proof proceeds as follows. First, we assume that $\tau = \{\mathsf{nop}, \mathsf{inp}, \mathsf{set}\}$ and that A is a directed cycle and argue that the τ-solvability of a given ESSP atom (k, s) or a SSP atom (s, s') of A is decidable in polynomial time. Secondly, we argue that the presented algorithmic approach is applicable to directed paths, too. Thirdly, we show that the procedure introduced for $\{\mathsf{nop}, \mathsf{inp}, \mathsf{set}\}$ can be extended to $\{\mathsf{nop}, \mathsf{inp}, \mathsf{set}, \mathsf{used}\}$. By Lemma 2, this proves the claim for $\{\mathsf{nop}, \mathsf{out}, \mathsf{res}\}$ and $\{\mathsf{nop}, \mathsf{out}, \mathsf{res}, \mathsf{free}\}$, too. After that we investigate the case where $\tau = \{\mathsf{nop}, \mathsf{set}, \mathsf{res}\} \cup \omega$ with nonempty $\omega \subseteq \{\mathsf{inp}, \mathsf{out}, \mathsf{used}, \mathsf{free}\}$. We argue that it is sufficient to decide the $\{\mathsf{nop}, \mathsf{inp}, \mathsf{res}, \mathsf{set}\}$- and $\{\mathsf{nop}, \mathsf{res}, \mathsf{set}, \mathsf{used}\}$-solvability of A and that this is doable in polynomial time. The corresponding procedures again modify those introduced for $\{\mathsf{nop}, \mathsf{inp}, \mathsf{set}\}$.

Let $\tau = \{\mathsf{nop}, \mathsf{inp}, \mathsf{set}\}$ and A be 1-bounded (cycle) TS with event $k \in E(A)$ that occurs m times. Since A is a cycle, we can assume that k occurs at A's initial state: $\iota \xrightarrow{k}$. Moreover, since k does not occur twice in a row, its occurrences partition A into m k-free subsequences $I_0, \ldots, I_{m-1}$ such that

$$I_i = s_0^i \xrightarrow{y_1^i} s_1^i \ldots s_{n_i-1}^i \xrightarrow{y_{n_i}^i} s_{n_i}^i, \ i \in \{0, \ldots, m-1\}, \text{ and } s_{n_{m-1}}^{m-1} = \iota, \text{ cf. Fig. 6.}$$

Obviously, defining $sup(\iota) = 1$, $sig(k) = \mathsf{inp}$ and $sig(e) = \mathsf{set}$ for all $e \in E(A) \setminus \{k\}$ inductively yields a region (sup, sig) solving the ESSP atoms (k, s) where $\xrightarrow{k} s$. Thus, it remains to consider the case $\neg(\xrightarrow{k} s)$. Since $\neg(\xrightarrow{k} s)$, there is an $i \in \{0, \ldots, m-1\}$ such that s is a state of the i-th subsequence I_i. In particular, there is a $j \in \{1, \ldots, n_i - 1\}$ such that $s = s_j^i$. The state s_j^i divides I_i into the sequences $I_i^0 = s_0^i \xrightarrow{y_1^i} \ldots \xrightarrow{y_j^i} s_j^i$ and $I_i^1 = s_j^i \xrightarrow{y_{j+1}^i} \ldots \xrightarrow{y_{n_i}^i} s_{n_i}^i$, cf. Fig. 6.

If (sup, sig) is a region that solves α then $sig(k) = \mathsf{inp}$ and $sup(s_j^i) = 0$ is true. This implies for all $\ell \in \{0, \ldots, m-1\}$ that $sup(s_0^\ell) = 0$ and $sup(s_{n_\ell}^\ell) = 1$. Thus, it remains to define the signature of the events of $\bigcup_{\ell=0}^{m-1} E(I_\ell)$ such that

$$0 \xrightarrow{sig(y_1^\ell)} \ldots \xrightarrow{sig(y_{n_\ell}^\ell)} 1, \text{ for all } \ell \in \{0, \ldots, m-1\} \setminus \{i\}, \text{ and } 0 \xrightarrow{sig(y_1^i)} \ldots \xrightarrow{sig(y_j^i)} 0$$

and $0 \xrightarrow{sig(y_{j+1}^i)} \ldots \xrightarrow{sig(y_{n_i}^i)} 1$.

If there is for all $\ell \in \{0, \ldots, m-1\} \setminus \{i\}$ an event $e_\ell \in E(I_\ell)$ such that $e_\ell \notin E(I_i^0)$ and if there is an event $e_i \in E(I_i^1)$ so that $e_i \notin E(I_i^0)$ then $sup(\iota) = 1$,

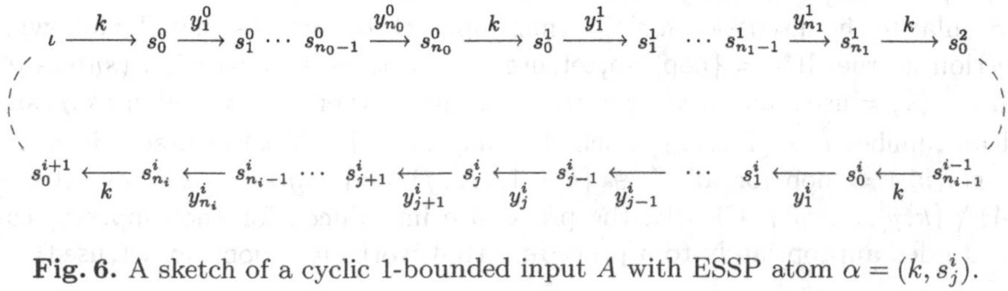

Fig. 6. A sketch of a cyclic 1-bounded input A with ESSP atom $\alpha = (k, s_j^i)$.

$sig(k) = \mathsf{inp}$, $sig(e_\ell) = \mathsf{set}$ for all $\ell \in \{0, \ldots, m-1\}$, and $sig(e) = \mathsf{nop}$ for all $e \in E(A) \setminus \{k, e_0, \ldots, e_\ell\}$ yields a τ-region (sup, sig) of A that solves α. Clearly, whether A satisfies this property is decidable in polynomial time.

Otherwise, there is a sequence $I \in \{I_0, \ldots, I_{i-1}, I_i^1, I_{i+1}, \ldots, I_{m-1}\}$ so that $E(I) \subseteq E(I_i^0)$. Thus, if (sup, sig) is a τ-region that solves α then there is a $\ell \in \{1, \ldots, j-1\}$ such that $sig(y_\ell^i) = \mathsf{set}$. Consequently, there has to be a $\ell' \in \{\ell + 1, \ldots, j\}$ such that $sig(y_{\ell'}^i) = \mathsf{inp}$ and, in particular, $sig(y_{\ell''}^i) = \mathsf{nop}$ for all $\ell'' \in \{\ell' + 1, \ldots, j\}$. Using this, one finds that (sup, sig) implies a region (sup', sig') that solves α and gets along with at most two inp-events. More exactly, defining $sup'(\iota) = 1$, $sig'(k) = sig'(y_{\ell'}^i) = \mathsf{inp}$, $sig'(e) = \mathsf{nop}$ for all $e \in \{y_{\ell'+1}^i, \ldots, y_j^i\}$ and $sig'(e) = \mathsf{set}$ for all $e \in E(A) \setminus (\{k, y_{\ell'}^i, \ldots, y_j^i\})$ yields a valid τ-region (sup', sig') that solves α. Since (sup, sig) was arbitrary, these deliberations show that in the second case the atom α is τ-solvable if and only if there is a corresponding region (sup', sig'). This yields the following polynomial procedure that decides whether α is τ-solvable: For ℓ from j to 2 test if (sup_ℓ, sig_ℓ) (inductively) defined by $sup_\ell(\iota) = 1$, $sig_\ell(y_\ell^i) = \mathsf{inp}$, $sig_\ell(y_{\ell'}^i) = \mathsf{nop}$ for all $\ell' \in \{\ell + 1, \ldots, j\}$ and $sig_\ell(e) = \mathsf{set}$ for all $e \in E(A) \setminus (\{k, y_\ell^i, \ldots, y_j^i\})$ yields a τ-region of A. If the test succeeds for any iteration then return yes, otherwise return no.

We can modify this approach to test the τ-solvability of an SSP atom $\beta = (s, s')$ as follows. Since $A = \iota \xrightarrow{e_1} \ldots \xrightarrow{e_m} \iota$ is a cycle we can assume without loss of generality that $s = \iota$ and $s' = s_i$ for some $i \in \{1, \ldots, m-1\}$. The states ι and s_i partition A into two subsequences $I_0 = \iota \xrightarrow{e_1} \ldots \xrightarrow{e_i} s_i$ and $I_1 = s_i \xrightarrow{e_{i+1}} \ldots \xrightarrow{e_m} \iota$. If β is solvable by a region (sup', sig') such that $sup'(\iota) = 1$ and $sup'(s_i) = 0$ then there is an event $e \in I_0$ such that $sig(e) = \mathsf{inp}$. In particular, there is a region (sup, sig) as follows: $sup(\iota) = 1$, $sig(e_j) = \mathsf{inp}$ and $j \in \{1, \ldots, i\}$, $sig(e_\ell) = \mathsf{nop}$ for all $\ell \in \{j+1, \ldots, i\}$ and $sig(e) = \mathsf{set}$ for all $e \in E(A) \setminus \{e_j, \ldots, e_i\}$. Similar to the approach for α, we can check if such a region exists in polynomial time. Moreover, the case where $sup(\iota) = 0$ and $sup(s_i) = 1$ works symmetrically.

So far we have shown that the τ-solvability of (E)SSP atoms of A are decidable in polynomial time if A is a cycle. If $A = \iota \xrightarrow{e_1} \ldots \xrightarrow{e_m} s_m$ is a directed path then its *cycle extension* A_c has a fresh event $\oplus \notin E(A)$ and is defined by $A_c = \iota \xrightarrow{e_1} \ldots \xrightarrow{e_m} s_m \xrightarrow{\oplus} \iota$. The event $\oplus$ is unique thus an (E)SSP atom of A is solvable by a τ-region of A if and only if it is solvable by a τ-region of A_c. Thus, we can decide the solvability of atoms of A via A_c. Altogether, this proves that the τ-solvability of (E)SSP atoms of 1-bounded inputs is decidable in polynomial time. Since we have at most $|S|^2 + |E| \cdot |S|$ atoms to solve, the decidability of the $\{\mathsf{nop}, \mathsf{inp}, \mathsf{set}\}$-solvability for 1-bounded TS is polynomial.

Similar to the discussion for $\tau = \{\mathsf{nop}, \mathsf{inp}, \mathsf{set}\}$, one argues that the following assertion is true: If $\tau = \{\mathsf{nop}, \mathsf{inp}, \mathsf{set}, \mathsf{used}\}$ then there is a τ-region (sup', sig') with $sig'(k) = \mathsf{used}$ and $sup'(s_j^i) = 0$ if and only if there is a τ-region (sup, sig) and an number $\ell \in \{1, \ldots, j\}$ such that $sup(\iota) = 1$, $sig(k) = \mathsf{used}$, $sig(y_\ell^i) = \mathsf{inp}$, $sig(y_{\ell'}^i) = \mathsf{nop}$ for all $\ell' \in \{\ell + 1, \ldots, j\}$ and $sig(e) = \mathsf{set}$ for all $e \in E(A) \setminus \{k, y_\ell^i, \ldots, y_j^i\}$. Clearly, the procedure introduced for $\{\mathsf{nop}, \mathsf{inp}, \mathsf{set}\}$ can be extended appropriately to a procedure that works for $\{\mathsf{nop}, \mathsf{inp}, \mathsf{set}, \mathsf{used}\}$.

It remains to investigate the case where $\tau = \{\text{nop}, \text{res}, \text{set}\} \cup \omega$ with non-empty $\omega \subseteq \{\text{inp}, \text{out}, \text{used}, \text{free}\}$. For a start, let's argue that deciding the τ-solvability is equivalent to deciding the $\{\text{nop}, \text{inp}, \text{res}, \text{set}\}$-solvability or the $\{\text{nop}, \text{res}, \text{set}, \text{used}\}$-solvability of A. This can be seen as follows: If (sup, sig) is a region that solves an ESSP atom $\alpha = (k, s)$ such that $sig(k) = \text{inp}$ then there is a $\{\text{nop}, \text{inp}, \text{res}, \text{set}\}$-region (sup', sig') that solves (k, s), too. The region (sup', sig') originates from (sup, sig) by $sup' = sup$, $sig'(k) = \text{inp}$ and for all $e \in E(A) \setminus \{k\}$ by $sig'(e) = \text{nop}$ if $sig(e) \in \{\text{nop}, \text{used}, \text{free}\}$, $sig'(e) = \text{res}$ if $sig(e) \in \{\text{inp}, \text{res}\}$ and, finally, $sig'(e) = \text{set}$ if $sig(e) \in \{\text{out}, \text{set}\}$. Similarly, one argues that α is τ-solvable such that $sig(k) = \text{out}$ if and only if it is $\{\text{nop}, \text{out}, \text{res}, \text{set}\}$-solvable. Moreover, $\{\text{nop}, \text{inp}, \text{res}, \text{set}\}$ and $\{\text{nop}, \text{out}, \text{res}, \text{set}\}$ are isomorphic thus τ-solvability with inp or out reduces to $\{\text{nop}, \text{inp}, \text{res}, \text{set}\}$-solvability. Similarly, the τ-solvability with used or free reduces to $\{\text{nop}, \text{res}, \text{set}, \text{used}\}$-solvability. It is easy to see that the procedure introduced for $\{\text{nop}, \text{inp}, \text{set}\}$ can be extended to the types $\{\text{nop}, \text{inp}, \text{res}, \text{set}\}$ and $\{\text{nop}, \text{res}, \text{set}, \text{used}\}$. The only difference is that we now search for an event y_ℓ^i such that $sig(y_\ell^i) = \text{res}$ instead of $sig(y_\ell^i) = \text{inp}$.

Finally, we observe that a SSP atom $\beta = (s, s')$ is τ-solvable if and only if it is $\{\text{nop}, \text{res}, \text{set}\}$-solvable. The states s and s' induce again a partition I_0 and I_1 of A and we can adapt the approach above to $\{\text{nop}, \text{res}, \text{set}\}$. $\qquad\square$

Theorem 10. *For any fixed $g \in \mathbb{N}$, deciding whether a g-bounded TS A is τ-solvable is polynomial if one of the following conditions is true:*

1. $\tau = \{\text{nop}, \text{inp}, \text{free}\}$ or $\tau = \{\text{nop}, \text{inp}, \text{used}, \text{free}\}$ or $\tau = \{\text{nop}, \text{out}, \text{used}\}$ or $\tau = \{\text{nop}, \text{out}, \text{used}, \text{free}\}$ and $g < 2$.
2. $\tau = \{\text{nop}, \text{set}, \text{res}\} \cup \omega$ and $\emptyset \neq \omega \subseteq \{\text{used}, \text{free}\}$ and $g < 3$.
3. $\tau = \tau' \cup \omega$ and $\tau' \in \{\{\text{nop}, \text{set}, \text{swap}\}, \{\text{nop}, \text{res}, \text{swap}\}, \{\text{nop}, \text{res}, \text{set}, \text{swap}\}\}$ and $\emptyset \neq \omega \subseteq \{\text{used}, \text{free}\}$ and $g < 2$.
4. $\tau \in \{\{\text{nop}, \text{inp}\}, \{\text{nop}, \text{inp}, \text{used}\}, \{\text{nop}, \text{out}\}, \{\text{nop}, \text{out}, \text{free}\}\}$ or $\tau \in \mathcal{T} = \{\{\text{nop}, \text{set}, \text{swap}\}, \{\text{nop}, \text{res}, \text{swap}\}, \{\text{nop}, \text{set}, \text{res}\}, \{\text{nop}, \text{set}, \text{res}, \text{swap}\}\}$,

Proof. (1): It is easy to see that A is a loop, $A \cong s\xrightarrow{e}s$ or that A is cycle free, since there is an unsolvable SSP atom otherwise. Moreover, if an event e occurs twice consecutively, $s\xrightarrow{e}s'\xrightarrow{e}s''$, then (s, s') is not τ-solvable. Thus, for every $e \in E(A)$ there is a $s \in S(A)$ such that (e, s) has to be solved by $sig(e) = \text{inp}$ ($sig(e) = \text{out}$) and $sup(s) = 0$ ($sup(s) = 1$). If e occurs twice on the directed path A then such a region does not exist. On the other hand, A is τ-solvable if every event occurs exactly once. Consequently, A is τ-solvable if and only if it is 1-bounded and every event occurs exactly once.

(2): Since ESSP atoms of a τ-solvable input A are only solvable by used and free, we have that if $s\xrightarrow{e}s' \in A$ then $s'\xrightarrow{e}s'' \in A$. If $s = s'' \neq s'$ or if s, s', s'' are pairwise distinct then (s, s') is not τ-solvable. This implies $s'\xrightarrow{e}s'$. As a result, τ-solvable inputs have the shape

$$A = \iota \xrightarrow{\ e_0\ } s_1 \quad \cdots \quad s_{m-1} \xrightarrow{\ e_m\ } s_m$$

with loops e_1 at s_1 and e_m at s_m

Thus, if the *loop erasement* A' of A originates from A by erasing all loops $s \xrightarrow{e} s$, that is, $A' = \iota \xrightarrow{e_1} \ldots \xrightarrow{e_m} s_m$, then deciding the τ-solvability of A reduces to deciding if A' has the τ-SSP and if all ESSP atoms (e, s) with $\neg(\xrightarrow{e} s)$ of A' are τ-solvable. This is doable in polynomial time by the approach of Theorem 9.

(3): Since ESSP atoms of an input A are only solvable by used and free, if $s \xrightarrow{e} s'$ and $s \neq s'$ then $s' \xrightarrow{e}$. If $s \xrightarrow{e} s' \xrightarrow{e} s'' \xrightarrow{e} s''' \in A$ and s, s', s'', s''' are pairwise different, then the SSP atom (s', s''') is not τ-solvable. As a consequence, τ-solvable inputs can have at most 3 different states.

(4): Let $\tau \in \{\{\mathsf{nop}, \mathsf{inp}\}, \{\mathsf{nop}, \mathsf{inp}, \mathsf{used}\}\}$. If A is τ-solvable, then for all $e \in E(A)$ holds $\iota \xrightarrow{e}$. Otherwise, (e, ι) is not τ-solvable. Similarly, if $\tau \in \mathcal{T}$, then ESSP atoms are not τ-solvable thus, every event occurs at ι. A is g-bounded. This implies $|E(A)| \leq g$. Thus, A has at most $2 \cdot |\tau|^g$ τ-regions. Since g is fixed, τ-synthesis is polynomial by brut-force. By Lemma 2, the claim follows.

$\square$

5   Conclusion

In this paper, we fully characterize the computational complexity of nop-equipped Boolean Petri nets from g-bounded TS for any fixed $g \in \mathbb{N}$. Our results show that if τ-synthesis is hard then it remains hard even for low bounds g. Moreover, they also show that when g becomes very small, sometimes it makes the difference between hardness and tractability, cf. Fig. 1 §1–§3 and §9, but sometimes it does not, cf. Fig. 1 §4–§7. In this sense, the parameter g helps to recognize interactions that contribute to the power of a type. By Theorem 3 and Theorem 9, $\{\mathsf{nop}, \mathsf{inp}, \mathsf{set}\}$-synthesis is hard if $g \geq 2$ and tractable if $g < 2$, respectively. By Theorem 5, $\{\mathsf{nop}, \mathsf{inp}, \mathsf{set}, \mathsf{free}\}$-synthesis remains hard for all $g \geq 1$. Thus, if restricted to 1-bounded inputs then the test interaction free makes the difference between hardness and tractability of synthesis. Surprisingly enough, by Theorem 9, replacing free by used makes synthesis from 1-bounded TS tractable again. It remains future work, to characterize the computational complexity of synthesis for the remaining 128 types which do not contain nop. Moreover, since τ-synthesis generally remains hard even for (small) fixed g, the bound of a TS is ruled out for FPT-algorithms. Future work might be concerned with parameterizing τ-synthesis by the *dependence number* of the searched τ-net: If $N = (P, T, f, M_0)$ is a Boolean net, $p \in P$ and if the *dependence number* d_p of p is defined by $d_p = |\{t \in T \mid f(p, t) \neq \mathsf{nop}\}|$ then the *dependence number* d of N is defined by $d = \max\{d_p \mid p \in P\}$. At first glance, d appears to be a promising parameter for FPT-approaches because this parameterization puts the problem into the complexity class XP: Since a τ-region of $A = (S, E, \delta, \iota)$ is determined by $sup(\iota)$ and sig, for each (E)SSP atom α there are at most $2 \cdot |\tau|^d \cdot \sum_{i=0}^{d} \binom{|E|}{i}$ fitting τ-regions solving α. Thus, by $|\tau| \leq 8$, τ-synthesis parameterized by d is decidable in $\mathcal{O}(|E|^d \cdot |S| \cdot \max\{|S|, |E|\})$.

References

1. Badouel, E., Bernardinello, L., Darondeau, P.: The synthesis problem for elementary net systems is NP-complete. Theor. Comput. Sci. **186**(1–2), 107–134 (1997)
2. Badouel, E., Bernardinello, L., Darondeau, P.: Petri Net Synthesis. Texts in Theoretical Computer Science. An EATCS Series, Springer (2015)
3. Badouel, E., Darondeau, P.: Trace nets and process automata. Acta Inf. **32**(7), 647–679 (1995)
4. Cortadella, J.: Private correspondance (2017)
5. Kleijn, J., Koutny, M., Pietkiewicz-Koutny, M., Rozenberg, G.: Step semantics of Boolean nets. Acta Inf. **50**(1), 15–39 (2013)
6. Montanari, U., Rossi, F.: Contextual nets. Acta Inf. **32**(6), 545–596 (1995)
7. Moore, C., Robson, J.M.: Hard tiling problems with simple tiles. Discrete Comput. Geom. **26**(4), 573–590 (2001)
8. Pietkiewicz-Koutny, M.: Transition systems of elementary net systems with inhibitor arcs. In: Azéma, P., Balbo, G. (eds.) ICATPN 1997. LNCS, vol. 1248, pp. 310–327. Springer, Heidelberg (1997). https://doi.org/10.1007/3-540-63139-9_43
9. Rozenberg, G., Engelfriet, J.: Elementary net systems. In: Petri Nets. Lecture Notes in Computer Science, vol. 1491, pp. 12–121. Springer (1996)
10. Schmitt, V.: Flip-flop nets. In: Puech, C., Reischuk, R. (eds.) STACS 1996. LNCS, vol. 1046, pp. 515–528. Springer, Heidelberg (1996). https://doi.org/10.1007/3-540-60922-9_42
11. Tredup, R.: The complexity of synthesizing nopequipped boolean nets from g-bounded inputs (technical report), to appear in CoRR (2019)
12. Tredup, R.: Hardness results for the synthesis of b-bounded Petri Nets. In: Donatelli, S., Haar, S. (eds.) PETRI NETS 2019. LNCS, vol. 11522, pp. 127–147. Springer, Cham (2019). https://doi.org/10.1007/978-3-030-21571-2_9
13. Tredup, R.: Tracking down the bad guys: reset and set make feasibility for flip-flop net derivatives NP-complete. ICE. EPTCS **304**, 20–37 (2019)
14. Tredup, R., Rosenke, C.: Narrowing down the hardness barrier of synthesizing elementary net systems. In: CONCUR. LIPIcs, vol. 118, pp. 16:1–16:15. Schloss Dagstuhl - Leibniz-Zentrum fuer Informatik (2018)
15. Tredup, R., Rosenke, C.: The complexity of synthesis for 43 Boolean Petri Net types. In: Gopal, T.V., Watada, J. (eds.) TAMC 2019. LNCS, vol. 11436, pp. 615–634. Springer, Cham (2019). https://doi.org/10.1007/978-3-030-14812-6_38
16. Tredup, R., Rosenke, C.: On the hardness of synthesizing Boolean nets. In: ATAED@Petri Nets/ACSD. CEUR Workshop Proceedings, vol. 2371, pp. 71–86. CEUR-WS.org (2019)
17. Tredup, R., Rosenke, C., Wolf, K.: Elementary net synthesis remains NP-complete even for extremely simple inputs. In: Khomenko, V., Roux, O.H. (eds.) PETRI NETS 2018. LNCS, vol. 10877, pp. 40–59. Springer, Cham (2018). https://doi.org/10.1007/978-3-319-91268-4_3

A Two-Player Asynchronous Game on Fully Observable Petri Nets

Federica Adobbati, Luca Bernardinello[✉], and Lucia Pomello

DISCo, Università degli Studi di Milano - Bicocca, Viale Sarca 336 U14, Milan, Italy
luca.bernardinello@unimib.it

Abstract. A Petri net is distributed if its elements can be assigned to a set of locations so that each element belongs to exactly one location, and each transition belongs to the same location as its input places.

We define an asynchronous game played on the unfolding of a distributed net with two locations, the 'user' and the 'environment'. The user can control the transitions in its location. A play in the game is a run in the unfolding, together with a sequence of cuts in that run. The rules of the game require that the environment satisfies a progress constraint: no transition in its location can be indefinitely postponed. In the general case, the game can be defined so that the user can observe only some transitions. In this paper, we only consider the case in which all transitions are observable, and study a reachability problem, in which the user tries to fire a target transition. We propose an algorithm which decides if the user has a winning strategy and, if so, computes a winning strategy.

1 Introduction

The ideas behind this paper were conceived while studying the problem of *weak observable liveness* [3,6], where we suppose that a Petri net models a system comprising a user and an environment; the user controls a subset of transitions, and observes a subset of transitions. The aim of the user is to force liveness of a special transition (the *target*), whatever the behaviour of the environment. The environment is supposed to guarantee progress of uncontrollable transitions.

The problem can be stated as deciding whether the user has a strategy allowing him to achieve his aim, irrespective of the choices of the environment. The strategy is formalized as a *response function*, mapping observations (sequences of observable transitions) to sets of controllable transitions.

In a first attempt to develop an algorithm for finding a strategy, the problem was translated into an infinite game on a finite graph, where the finite graph is derived from the marking graph of the net [3]. Besides the usual problem of state explosion, this approach hides the potential concurrency in the net, by using an interleaving semantics.

Hence, the authors started to explore the idea of defining an asynchronous game, to be played on the unfoldings of Petri nets, in which to encode the

© Springer-Verlag GmbH Germany, part of Springer Nature 2021
M. Koutny et al. (Eds.): ToPNoC XV, LNCS 12530, pp. 126–149, 2021.
https://doi.org/10.1007/978-3-662-63079-2_6

weak observable liveness problem, but also several other problems, formalized by defining a suitable aim for the user. Such a game was proposed in [4], where its application to weak observable liveness was studied.

Other possible applications of such a game could be in the general frame of verification, adaptation and control of distributed systems; so that, in the case of a winning strategy for the user with respect to a specific behavioral property, the system model could be adapted imposing that specific property, for example by adding an interacting component which implements the user behavior by synthesising the winning strategy; a reference for this sort of applications could be for example [12].

In this paper, we consider *distributed* net systems, in which all choices are local to one component, restricted to the case of exactly two components, user and environment, where the user has a sequential behaviour, whereas in the environment, transitions can be concurrent with each other, and with user's transitions.

We propose an asynchronous game played on the unfolding of the system in a general setting, so that by defining proper strategies, we can adapt the same model for the verification of different properties. Here we study a reachability problem, in which the user tries to fire a target transition.

In the general case, in which the user can observe the occurrence of only some environment's transitions, the definition of the game on unfolding may allow to define a winning strategy for the user, whereas this would be not possible by considering a game based on interleaving semantics. This fact has been briefly discussed on the basis of an example in [2] and is motivated by the fact that, in the unfolding, it is possible to distinguish different occurrences of the same transition, occurrences which can be differently related to other occurrences of another transition. In this way, the structure of the unfolding, even with partial observability, may allow to reconstruct the unobservable evolution of the system.

Obviously, the lack of information may even prevent to find a winning strategy; the chances of having a winning strategy for the user increase by observing as much as possible the behaviour of the environment; and if the user has no strategy by observing every transitions, there is no hope in the case of partial observability.

As a first step towards the identification of an algorithm in the general case of partial observability, in this paper we assume full observability, and propose an algorithm on the unfolding which decides if the user has a winning strategy and, if so, it computes such a winning strategy.

The paper, which is an extended, revised version of [1], is structured as follows. In the next section, we recall the needed notions about Petri nets, distributed Petri nets, and unfoldings, In Sect. 3 we define the general game, and the notions of strategy and winning strategy. The problem of controlled reachability is introduced in Sect. 4, together with the algorithm looking for a winning strategy. Several approaches to notions of asynchronous games are briefly discussed in Sect. 5, while prospects for future developments are presented in the final section.

2   Petri Nets

Petri nets model concurrent systems. The basic elements of a net are local states (places) and local transitions. The global state of a net is distributed among its local states. When a transition occurs, it changes the value of local states in its neighbourhood. Several types of nets have been defined and studied. Here, we use *1-safe* net systems.

Definition 1. *A net is a triple* $N = (P, T, F)$, *where* P *and* T *are disjoint sets. The elements of* P *are called* places *and represented by circles, the elements of* T *are called* transitions *and represented by squares.* F *is called* flow relation, *with* $F \subseteq (P \times T) \cup (T \times P)$, *and is represented by arcs.*

Let $x \in P \cup T$ be an element of the net; the pre-set of x is the set ${}^\bullet x = \{y \in P \cup T \mid (y, x) \in F\}$, the post-set of x is the set $x^\bullet = \{y \in P \cup T \mid (x, y) \in F\}$.

We assume that any transition has non-empty pre-set and post-set: $\forall t \in T$, ${}^\bullet t \neq \emptyset$ and $t^\bullet \neq \emptyset$.

A net is infinite if $P \cup T$ is infinite, finite otherwise.

Two transitions, t_1 and t_2, are *independent* if $({}^\bullet t_1 \cup t_1^\bullet)$ and $({}^\bullet t_2 \cup t_2^\bullet)$ are disjoint.

A net $N' = (P', T', F')$ is a *subnet* of $N = (P, T, F)$ if $P' \subseteq P$, $T', \subseteq T$, and F' is F restricted to the elements in N'.

Definition 2. *A net system is a quadruple* $\Sigma = (P, T, F, m_0)$ *consisting of a finite net* $N = (P, T, F)$ *and an* initial marking $m_0 : P \to \mathbf{N}$.

A transition t is *enabled* at a marking m, denoted $m[t\rangle$, if, for each p in ${}^\bullet t$, $m(p) > 0$. A transition t, enabled at m, can *occur* (or *fire*) producing a new marking

$$
m'(p) = \begin{cases} m(p) + 1 & \text{if } p \in t^\bullet \setminus {}^\bullet t \\ m(p) - 1 & \text{if } p \in {}^\bullet t \setminus t^\bullet \\ m(p) & \text{otherwise} \end{cases}
$$

A marking m' is reachable from another marking m, if there is a sequence $t_1 t_2 \ldots t_n$ such that $m[t_1\rangle m_1[t_2\rangle \ldots m_{n-1}[t_n\rangle m'$; in this case, we write $m' \in [m\rangle$. The set of reachable markings is the set of markings reachable from the initial marking m_0, denoted $[m_0\rangle$.

A net system is *1-safe* if, for each reachable marking m, and each place p, $m(p) \leq 1$. Markings in 1-safe net systems can, and will be, considered as subsets of places.

In a net system, two transitions, t_1 and t_2, are *concurrent* at a marking m if they are independent and both enabled at m.

The non sequential behaviour of 1-safe net systems can be recorded by occurrence nets, which are used to represent by a single object the set of potential histories of a net system. In the following, by F^* we denote the reflexive and transitive closure of F.

Two elements $x, y \in P \cup T$ are said to be in *conflict*, denoted $x \# y$, iff there exist $t_1, t_2 \in T : t_1 \neq t_2, t_1 F^* x, t_2 F^* y \wedge \exists p \in {}^\bullet t_1 \cap {}^\bullet t_2$.

Definition 3. *A net* $N = (B, E, F)$ *is an* occurrence net *if*

- *for all* $b \in B$, $|{}^\bullet b| \leq 1$
- F^* *is a partial order on* $B \cup E$
- *for all* $x \in B \cup E$, *the set* $\{y \in B \cup E \mid yF^*x\}$ *is finite*
- *for all* $x \in B \cup E$, $x \# x$ *does not hold*

We will say that two elements x and y, $x \neq y$, of N are *concurrent*, and write x **co** y, if they are not ordered by F^*, and they are not in conflict.

By $\min(N)$ we will denote the set of minimal elements with respect to the partial order induced by F^*.

A *B-cut* of N is a maximal set of pairwise concurrent elements of B. B-cuts represent potential global states through which a process can go in a history of the system. By analogy with net systems, we will sometimes say that an event e of an occurrence net is *enabled* at a B-cut γ, denoted $\gamma[e\rangle$, if ${}^\bullet e \subseteq \gamma$. We will denote by $\gamma + e$ the B-cut $(\gamma \setminus {}^\bullet e) \cup e^\bullet$. A B-cut is a *deadlock cut* if no event is enabled at it.

Let Γ be the set of B-cuts of N. A partial order on Γ can be defined as follows: let γ_1, γ_2 be two B-cuts. We say $\gamma_1 < \gamma_2$ iff

1. $\forall y \in \gamma_2 \exists x \in \gamma_1 : xF^*y$
2. $\forall x \in \gamma_1 \exists y \in \gamma_2 : xF^*y$
3. $\exists x \in \gamma_1, \exists y \in \gamma_2 : xF^+y$

In words, $\gamma_1 < \gamma_2$ if any condition in the second B-cut is or follows a condition of the first B-cut and any condition in the first B-cut is or comes before a condition of the second B-cut (and there exists at least one condition coming before).

A sequence of B-cuts, $\gamma_0 \gamma_1 \ldots \gamma_i \ldots$ is *increasing* if $\gamma_i < \gamma_{i+1}$ for all $i \geq 0$.

We will say that an event $e \in E$ precedes a B-cut γ, and write $e < \gamma$, iff there is $y \in \gamma$ such that eF^+y. In this case, each element of γ either follows e or is concurrent with e in the partial order induced by the occurrence net.

Definition 4. *A* branching process *of* $\Sigma = (P, T, F, m_0)$ *is an occurrence net* $N = (B, E, F)$, *together with a labelling function* $\mu : B \cup E \to P \cup T$, *such that*

- $\mu(B) \subseteq P$ *and* $\mu(E) \subseteq T$
- *for all* $e \in E$, *the restriction of* μ *to* ${}^\bullet e$ *is a bijection between* ${}^\bullet e$ *and* ${}^\bullet\mu(e)$; *the same holds for* $e^\bullet$
- *the restriction of* μ *to* $\min(N)$ *is a bijection between* $\min(N)$ *and* m_0
- *for all* $e_1, e_2 \in E$, *if* ${}^\bullet e_1 = {}^\bullet e_2$ *and* $\mu(e_1) = \mu(e_2)$, *then* $e_1 = e_2$

A *run* of Σ is a branching process (N, μ) such that the conflict relation of the underlying occurrence net is empty.

For γ a B-cut of N, the set $\{\mu(b) \mid b \in \gamma\}$ is a reachable marking of Σ, and we refer to it as the marking corresponding to γ.

Let (N_1, μ_1) and (N_2, μ_2) be two branching processes of Σ. We say that (N_1, μ_1) is a *prefix* of (N_2, μ_2) if N_1 is a subnet of N_2, and

- $\min(N_1) = \min(N_2)$
- if $b \in B_1$ and $(e, b) \in F_2$, then $e \in E_1$
- if $e \in E_1$, and b is either a precondition or a postcondition of e in N_2, then $b \in B_1$

For any 1-safe net system Σ, there exists a unique, up to isomorphism, maximal branching process of Σ. We will call it the *unfolding* of Σ, and denote it by UNF(Σ) (see [7]).

A *run* of Σ describes a particular history of Σ, in which conflicts have been solved. Any run of Σ is a prefix of the unfolding UNF(Σ); we will also say that it is a run on UNF(Σ).

In this paper we are interested in Petri nets modelling systems in which a *User* controls a subset of transitions, while interacting with an *Environment*. Intuitively, this means that the User can decide whether to fire such a transition when it is enabled.

We also assume that choices among transitions are local either to the Environment or to the User, and that transitions controlled by the User are never concurrent with each other, while they can be concurrent with transitions in the Environment.

As a formal setting, we refer to the so-called *distributed net systems*, as introduced and studied in [5] and in [10].

Definition 5. *A distributed net system over a set L of locations is a 1-safe net system $\Sigma = (P, T, F, m_0)$ together with a map*

$$\lambda : (P \cup T) \to L$$

such that for every $p \in P$, $t \in T$, if $p \in {}^{\bullet}t$, then $\lambda(p) = \lambda(t)$.

In this paper, we consider the special case of distributed net systems $\langle \Sigma, \lambda \rangle$ such that $L = \{\text{Environment}, \text{User}\}$, i.e. of distributed net systems with only two components, representing the Environment and the User, respectively; we assume that the User controls all transitions in its location, and these transitions are never concurrent with each other. From now on, by distributed net system we will mean a net system satisfying these constraints.

In distributed net systems, when a transition is enabled, it can never be disabled by the occurrence of transitions belonging to different components. In the case of a cycle this observation justifies the following lemma.

Lemma 1. *Let $\langle \Sigma, \lambda \rangle$ be a distributed net system with two locations, A and G. Let m be a marking, and*

$$m_1[t_1\rangle m_2[t_2\rangle m_3[...\rangle m_1$$

be a firing sequence with $\lambda(t_i) = A$ for each i. Then, if $\lambda(t) = G$, and t is enabled at m_i for some i between the two repetitions of m_1, then t is enabled at m_j for each m_j in the cycle.

The notions of unfolding and run apply in the obvious ways to distributed net systems. We will use E_c to denote the set of controllable events in the unfolding (occurrences of controllable transitions, performed by the *User*), and $E_{nc} = E \setminus E_c$ to denote uncontrollable events. Uncontrollable transitions are meant to represent actions performed by the *Environment*.

In the graphical representation, controllable transitions and events will be represented by black squares.

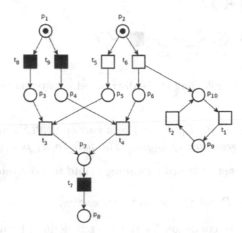

Fig. 1. A distributed net system with two locations

Example 1. Figure 1 shows a distributed net system with two locations. Places are not explicitly divided into the two components, because their partition can be inferred by their post-transitions. A prefix of the unfolding of the system is shown in Fig. 2. Each element of the unfolding is decorated with the label of an element in the net, with an exponent which distinguishes different occurrences of the same element. The dotted line suggests that the unfolding goes on by repeating occurrences of transitions t_1 and t_2, and of their neighbouring places.

3   An Asynchronous Game on the Unfolding

Let Σ be a distributed net system with two locations, Environment and User. We assume that the Environment is subject to a progress (or weak fairness) property: if an uncontrollable transition is enabled, then it will eventually either fire or become disabled.

We define a game on UNF(Σ). A play in the game is a run, weakly fair with respect to uncontrollable transitions, together with an increasing sequence of B-cuts, which can be seen as a potential record of the play as observed by an external entity. Several transitions can occur between two contiguous cuts in the sequence.

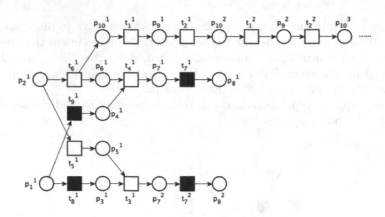

Fig. 2. The unfolding of the distributed net system in Fig. 1

Definition 6. *Let* $\rho = (B_\rho, E_\rho, F_\rho, \mu_\rho)$ *be a run on* UNF(Σ) *and* $\pi = \gamma_0, \gamma_1, \cdots,$ $\gamma_i, \cdots$ *an increasing sequence of B-cuts. The pair* (ρ, π) *is said to be a* play *if:*

- $\forall e \in E_{nc} \backslash E_\rho$, *the net obtained by adding* e *and its postconditions to* ρ *is not a run of* UNF(Σ);
- $\forall e \in E_\rho$ *there is a B-cut* $\gamma_i \in \pi$ *such that* $e < \gamma_i$.

In general, the winning condition for the User is defined by a set of plays. The significant cases to analyse are those in which the winning set is determined by a property that we are interested in investigating on the model.

For example, let us suppose that we are interested in knowing if a user is able to force the occurrence of a target transition once. We can model this problem as a game in which the User wins a play if the corresponding run contains an occurrence of the target transition.

Another possible goal of a play, as analysed for example in [9], is to verify if it is always possible to avoid a certain marking in a controllable system. In this case the User wins those plays in which there are no cuts associated with that marking. Whatever the goal of the game is, a strategy is a function formalizing the behaviour of the User during a play.

In general, one might suppose that the User cannot observe everything in the system. For instance, it might not directly observe firings of some transitions in the Environment. In this paper, we suppose that the User can see all occurrences of transitions. This implies that the User can determine the current cut in the unfolding on the basis of the transition occurrences observed so far; hence, a strategy can be defined as a map from B-cuts to sets of controllable events.

Definition 7. *Let* Γ *be the set of B-cuts in* UNF(Σ). *A strategy is a function* $\alpha : \Gamma \to 2^{E_c}$ *such that for every* $\gamma \in \Gamma$ *and for every* $e \in E_c$, *if* $e \in \alpha(\gamma)$, *then* e *is enabled in* γ.

Definition 8. *Let* (ρ, π) *be a play. An event* $e \in E_c$ *is* finally postponed *in* (ρ, π) *iff there is a cut* $\gamma_i \in \pi$ *in which* e *is enabled and such that* $\forall k \geq i, \gamma_k[e\rangle$.

Definition 9. *Let (ρ, π) be a play and α be a strategy. An event $e \in E_c$ is finally eligible in (ρ, π) by α iff there is a cut $\gamma_i \in \pi$ such that $e \in \alpha(\gamma_i)$ and $\forall k \geq i$, $e \in \alpha(\gamma_k)$.*

A play complies with a strategy if all controllable events in the play have been chosen according to the strategy, and no controllable event is finally postponed and eligible.

Definition 10. *Let $\rho = (B_\rho, E_\rho, F_\rho, \mu_\rho)$ be a run in $\mathrm{UNF}(\Sigma)$, $\pi = \gamma_1 \gamma_2, \dots$ be an increasing infinite sequence of B-cuts and α be a strategy. The pair (ρ, π) is an α−play iff:*

1. *(ρ, π) is a play;*
2. *For every controllable event e belonging to E_ρ, there must be a B-cut $\gamma_i \in \pi$ such that $e \in \alpha(\gamma_i)$ and $\gamma_{i+1} = (\gamma_i \backslash {}^\bullet e) \cup e^\bullet$.*
3. *If $|E_\rho \cap E_c| < \infty$, there is no event $e \in E_c \cap E_\rho$ finally eligible by α and finally postponed in the play.*

A strategy $\alpha : \Gamma \to 2^{E_c}$ is winning iff the User wins all the α-plays. In general, if there is a winning strategy, it is not unique.

Example 2. The net system shown in Fig. 1 is distributed, with two locations. Define a game on its unfolding, shown in Fig. 2, so that the User wins a play if the play contains an occurrence of t_7.

By inspecting the net, it is clear that a winning strategy for the User consists in waiting for the Environment to choose between t_5 and t_6, and then fire, respectively, either t_8 or t_9. Since the Environment cannot postpone its choice forever, and will be forced to eventually fire either t_3 or t_4, the User will be able to fire t_7, and win the game. Formally, the winning strategy can be defined as follows: $\alpha(\{p_1^1, p_6^1, p\}) = \{t_9^1\}$, where p is any occurrence of either p_9 or p_{10}, $\alpha(\{p_1^1, p_5^1\}) = \{t_8^1\}$, $\alpha(\{p_7^2\}) = \{t_7^2\}$, $\alpha(\{p_7^1, p\}) = \{t_7^1\}$, where p is any occurrence of either p_9 or p_{10}, $\alpha(\gamma) = \emptyset$ for any other B-cut γ. In particular, $\alpha(\{p_1^1, p_2^1\}) = \emptyset$, to encode the decision to wait, in the initial cut, for the Environment to choose its first move. Figure 3 shows an α-play.

4   Controlled Reachability

In this section we apply the general idea of asynchronous game to a specific reachability problem, and propose an algorithm to determine if the User has a winning strategy.

Let $\langle \Sigma, \lambda \rangle$, where $\Sigma = (P, T, F, m_0)$, be a distributed net system. The problem of *controlled reachability* consists in determining if the User is able to lead the system to fire a certain transition once, despite the Environment behaviour, starting from the initial marking. This can be analysed through a game on the unfolding. Let t be the target transition; we define as winning condition for the User the set of plays (ρ, π) in which there is an event $e \in E_\rho$ labelled with t.

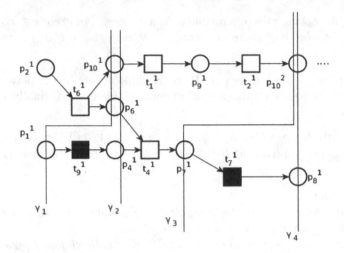

Fig. 3. An α-play on the unfolding in Fig. 2

A target transition t is controllably reachable in Σ if, and only if, there is a strategy α on UNF(Σ) such that the User wins every α-play. Example 2 above can be seen as a game of controlled reachability. The strategy discussed in the example is a winning strategy for this game.

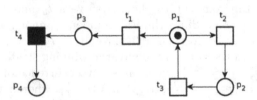

Fig. 4. A distributed net system

Example 3. The net shown in Fig. 4 is distributed, with two locations. Consider the game of controlled reachability played on its unfolding, shown in Fig. 5, where the target transition is t_4. If the Environment cooperates with the User by eventually choosing t_1, then the target is reached. However, the Environment can choose t_2 at every cut consisting in an occurrence of p_1. The Environment is subject to a weak fairness constraint, but not to a strong fairness constraint. Hence, irrespective of the strategy chosen by the User, an infinite play made of repeated occurrences of the cycle p_1, t_2, p_2, t_3, p_1 is admissible.

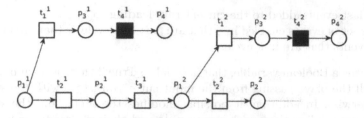

Fig. 5. The (prefix of the) unfolding of the net system shown in Fig. 4

In a general case, given a strategy α, there are infinitely many $\alpha-$plays in UNF(Σ), and some plays could be infinite, hence the exhaustive exploration of them would take infinite time. We propose an algorithm that, given a distributed net system and a target transition, establishes if there is a winning strategy for the controlled reachability of the target and, if so, computes a winning strategy.

4.1   Algorithm for a Winning Strategy

In this section we present the algorithm looking for a winning strategy for the reachability of a target transition on fully observable systems, and we illustrate it on the system in Fig. 1, already discussed in Example 2. The algorithm we present generates a prefix of the unfolding of a given net system, deciding whether there exists a winning strategy for the User. In the positive case, it gives as output a strategy as a function on reachable markings; the strategy is initially associated to B-cuts of the unfolding, but the algorithm works so that, for distinct cuts corresponding to the same marking, the strategy gives the same answer.

The input data are the following:

- A net in which the transitions are enumerated so that all the uncontrollable transitions precede all the controllable ones. If the target is a controllable transition, it must be the first of the controllable transitions.
- The position of the first controllable transition.
- The initial marking m_0 of the system.
- The target transition.

The value of these variables is available for all the functions of the algorithm and does not change during its execution.

The core of the algorithm is the recursive function UNF_EXPLORATION (see Algorithm 1), which unfolds the net by exploring reachable cuts, and constructs at the same time a prefix of the unfolding and a strategy.

The function takes five input arguments:

1. γ: the cut that must be analysed;
2. M: the list of markings associated to the cuts already analysed in the current run;
3. E_l: the list of events that fired in the current run;

4. e: the last event added to the current run, leading to γ;
5. sz: the set of events enabled in γ that are part of a cycle or that are in conflict with events that are in a cycle.

It returns a Boolean variable, that is equal to True if there is a winning strategy, for all the plays passing from the input cut γ consistent with the strategy, False otherwise. In addition, it possibly modifies the prefix and the strategy, initially empty, filling them with events, cuts and choices already explored.

The first time that the function is called, the input consists always in the initial cut γ_0 in the unfolding, empty lists for the list of visited markings, the list of analysed events and the list of events that are part of cycles or in conflict with them (those events will be discovered during the execution of the algorithm), a fictitious event i. The function UNF_EXPLORATION uses some auxiliary functions:

- ENAB_N is a function that has an input cut and returns the list of uncontrollable events which are enabled in that specific cut;
- similarly, ENAB_C returns the controllable enabled events.
- EXTRACT returns the first element of an input list, and the list deprived of this element.
- STABLE_ZONE returns the set of events that can be part of a cycle, and those in conflict with them.

Let us recall that we denote with $\gamma + e$ the cut obtained by firing the event e in the cut γ.

The function constructs every run by adding uncontrollable events until one of the following cases occurs: (1) the target occurs; (2) a deadlock cut is reached or a cut is reached in which only transitions that are part of a cycle or that are in conflict with events in a cycle are enabled; (3) a cut that has been previously analysed is reached (two subcases are considered); (4) a cut is reached in which no uncontrollable event is enabled, and some controllable events are enabled; (5) a cut is reached corresponding to a marking which has already been visited in the current run, and there are not uncontrollable enabled events that are concurrent with all the ones that occurred between the two equivalent markings.

In case (1), the current run corresponds to a play won by the User; hence the function tries to backtrack along choices among uncontrollable events, if possible. Symmetrically, in case (2), the current run corresponds to a play won by the Environment; hence, the function tries to backtrack along choices among controllable events, if possible. In cases (3), the current run is the prefix of a set of runs that have been already analysed. The User wins or loses according to the analysis previously done. In case (4), a controllable event is added, and the exploration restarts from the new cut. Finally, in case (5), if possible, a controllable event is added, and the exploration restarts from the new cut; if this is not possible, the run corresponds to a play won by the Environment and the function tries to backtrack and change the previous controllable choices.

Example 4. Consider the net system shown in Fig. 1, and its unfolding (Fig. 2), where the ordering on the set of transitions is given by their indices. Starting from

Algorithm 1. Unfolding exploration

```
function UNF_EXPLORATION(γ, M, E_l, e, sz)
    if e == target then return true
    else if γ is a deadlock or enables only transitions in sz then return false
    else if γ ∈ Γ_bad then return false
    else if γ ∈ Γ_good then return true
    else if μ(γ) ∈ M then return EXPLORE_CUT_C(γ, M, E_l)
    else if ENAB_N(γ) ≠ ∅ then
        E = ENAB_N(γ)
        repeat
            e_0, E = EXTRACT(E)
            v = UNF_EXPLORATION(γ + e_0, M.append(μ(γ)), E.append(e), e_0)
            if v == true then
                unf = unf ∪[γ, e_0, γ + e_0]
            end if
        until E == ∅ ∨ v == false
        if v == true then
            if γ ∈ ver then
                sz = STABLE_ZONE(E)
                v = UNF_EXPLORATION(γ, M, E_l, e)
            else
                Γ_good.append(γ)
            end if
        else
            Γ_bad.append(γ)
        end if
        return v
    else
        E = ENAB_C(γ)
        repeat
            e_0, E = EXTRACT(E)
            v = UNF_EXPLORATION(γ + e_0, M.append(μ(γ)), E_l.append(e), e_0)
            if v == true then
                unf = unf ∪[γ, e_0, γ + e_0]
                str = str ∪ [γ, e_0]
            end if
        until E == ∅ ∨ v == true
        if v == true then
            if γ ∈ ver then
                sz = STABLE_ZONE(E)
                v = UNF_EXPLORATION(γ, M, E_l, e)
            else
                Γ_good.append(γ)
            end if
        else
            Γ_bad.append(γ)
        end if
        return v
    end if
end function
```

the initial B-cut, the algorithm adds the event t_5^1, reaching a cut in which only controllable transitions are enabled. It then adds t_8^1, reaching the cut $\{p_3^1, p_5^1\}$, and starts again adding uncontrollable transitions. This run will lead to the target event t_7^2, hence it is not necessary to backtrack on controllable events.

The next backtracking step goes back to the initial cut, and starts exploring a new run by adding t_6^1; from $\{p_1^1, p_6^1, p_{10}^1\}$, the events t_1^1 and t_2^1 fire. This produces the cut $\{p_1^1, p_6^1, p_{10}^2\}$, that corresponds to a marking that has already been visited. Hence, the controllable event t_8^1 is added, leading to a cut in which only the cycle formed by occurrences of t_1 and t_2 can occur, thus repeating the same marking. The algorithm backtracks and tries t_9^1. The events t_1^2 and t_4^1 are enabled in $\{p_4^1, p_6^1, p_{10}^1\}$. Due to the order of the transitions of the net, t_1^2 and t_2^2 occur, reproducing the same marking. In order to guarantee the progress of the system, the algorithm adds only events that have been enabled since the first repetition of the marking associated with the current cut and have never been disabled from that moment on.

In Example 4, the only event that satisfies these requirements is t_4^1. By proceeding in this way, the algorithm continues until the target is reached.

In the following, we explain in detail how UNF_EXPLORATION works in a general step of execution of the algorithm. If γ is a cut of a play on the unfolding, one of these situations is verified:

1. γ is not a deadlock, it enables events that are not part of cycles or in conflict with them, has not been previously analysed, it is the first time that the associated marking is visited in the play, the target has not occurred yet and there are k uncontrollable enabled transitions to analyse in $\mu(\gamma)$. In this case, the prefix of the play currently ending with γ is extended in k different plays, each of them obtained by adding a different uncontrollable event after γ. The output for this step is True only if the values returned by all recursive calls on the cuts that immediately follow γ is True.
   Considering the system in Fig. 1 and its unfolding (Fig. 2), we find the described situation in the initial cut of the unfolding: in $\{p_1^1, p_2^1\}$, both the events t_6^1 and t_5^1 are enabled. Therefore, the algorithm extends the current prefix considering the two plays obtained by adding the two events and the cuts that follow their occurrence.

2. γ is not a deadlock, $\mu(\gamma)$ has never been analysed in the play, the target did not fire in the previous part of the play and the only enabled events that are not part of cycles or in conflict with them are controllable. In this case, the algorithm analyses the controllable events in the order induced by enumeration of the transitions in the net, until it finds an extension that returns True as output or it ends the analysis of all the controllable events enabled in γ.
   Referring to Fig. 2, the cut $\{p_1^1, p_5^1\}$ enables t_8^1 and t_9^1. The algorithm starts constructing the play with t_8^1. After verifying that the User wins all the α-plays passing from the cut $\{p_3^1, p_5^1\}$, the function does not continue with the analysis of t_9^1, and returns the Boolean value True.

3. Either γ is a deadlock, or all the enabled events are part of a cycle or in conflict with events in a cycle, or γ follows the target transition. These are base cases for the recursive algorithm. Their occurrence stops the exploration for that play. If the target fired, the algorithm returns True, in all the other cases of this situation, it returns False.

   In the considered example, all the plays ending with a cut in which there is an occurrence of p_8 are winning for the User (because an occurrence of t_7 has necessarily fired).

4. γ has already been considered in a previous step. In this case, the analysis stops and the function returns True, if the first analysis of the cut returned True, and False otherwise. This case is verified in case of concurrency in the Environment component.

5. $\mu(\gamma)$ was already visited in the play. In this case, the algorithm checks if any controllable event fired between the two repetitions. If this happens it returns False. (This is justified by the fact that the victory of the user cannot depend on the choice of a controllable transition that contributes to a cycle without the target.) Otherwise, it analyses only the events that are enabled and concurrent with all the ones fired in the cycle. If there are uncontrollable events among them, then it behaves as in 1; if there are only controllable events, it behaves as in 2; if there is no event satisfying the requirements, it behaves like in a deadlock situation.

   During the execution of the algorithm on the system in Fig. 1, the cut $\{p_3^1, p_6^1, p_{10}^3\}$ is analysed. The only enabled event is t_1^3, but it is not added to the play, because it depends on the repeated occurrences of transitions t_1 and t_2, that create a cycle in the system. Hence, the algorithm returns False for this particular play. Later, changing the controllable choice, it analyses the cut $\{p_4^1, p_6^1, p_{10}^3\}$. In this cut, t_4^1 is enabled and concurrent with all the occurrences of t_1 and t_2, hence, the algorithm extends the play with it.

The functions EXPLORE_CUT_C (Algorithm 2) and F (Algorithm 3) deal with concurrency. Specifically, EXPLORE_CUT_C is called by UNF_EXPLORATION when a cut associated with a marking repeated in the run is detected. The function F is called by EXPLORE_CUT_C; it takes the current cut, the list of the previously visited markings, and the list of events that have been fired. It checks whether a controllable event fired in the cycle; if not, it returns the list E of events concurrent with all the events occurred after the first cut in the run associated to the same marking as the current one. The events in E are the only ones considered by EXPLORE_CUT_C to extend the prefix of the run.

In Algorithm 2, there are two more auxiliary functions:

- E_{NC} takes a list of events as input, and returns only the uncontrollable ones.
- Symmetrically, E_C takes a list as input, and returns the controllable events in it.

Both UNF_EXPLORATION and EXPLORE_CUT_C are responsible for the construction of the prefix and the strategy. The prefix is updated every time that UNF_EXPLORATION returns the value True (with the exception of the very first call). When this happens, the receiving function appends to the prefix a triple consisting of its input cut γ, the following cut $\gamma + e$ that was in input to the call to the function that just returned True, and the event e. If the added event e is controllable, then the strategy is also updated. In particular, the algorithm appends the input cut γ coupled with the controllable transition $\mu(e)$ to the current strategy.

At the end of the execution of UNF_EXPLORATION, if there is a winning strategy, it is defined on the cuts of the prefix. To complete it, we have to define it on the markings, detect the parts of the plays corresponding to a cyclic behaviour on the system and, if the strategy chooses a transition immediately after them, the algorithm has to fill the strategy, attributing the same choice to all the markings in the cycle.

4.2 Discussion

In this section, we discuss the correctness of the proposed algorithm.

Lemma 2. *Every play exploration ends due to one of the following ending criteria:*

1. *The target fires. In this case the User wins all the α−plays with the constructed prefix.*
2. *The play reaches a deadlock cut γ before reaching the target. In this case the User loses the play.*
3. *The play reaches a cut in which the target has not fired, and the only enabled transitions can be part of cycles or in conflict with transitions that can be part of a cycle. In this case the user loses the play.*
4. *The play reaches a cut γ that was previously analysed.*
5. *The play reaches a cut γ' such that there is another cut $\gamma : \gamma < \gamma'$ for which $\mu(\gamma) = \mu(\gamma')$, γ corresponds to the first occurrence of $\mu(\gamma)$ in the play, and*
   − *either γ' does not enable any event that is concurrent with all the events occurred between γ and γ',*
   − *or there is a controllable event e such that $\gamma < e < \gamma'$.*
   If the prefix is consistent with the strategy, the User loses at least an α−play.

Moreover, if α is a strategy defined on the markings, then, for every prefix of an α−play determined with one of these criteria, we can decide if the User wins all the α−plays starting with such a prefix.

Proof. 1. If the target fired in the prefix, then every play with such a prefix is winning for the User, because it includes the target.
2. If the target does not fire and the play is in a deadlock, the prefix coincides with the whole play. Since it does not have the target, it is losing for the User.
3. If the target does not fire and the only enabled transitions can be part of a cycle or in conflict with transitions in cycles, then the user cannot prevent the environment to remain in the cycles forever (the transitions in a cycle are uncontrollable by construction). Since the target is not part of this cycle, the user cannot be sure to reach it.

Algorithm 2. Cuts associated with markings already visited in the prefix

Input: the cut γ that must be analysed, the ordered list M and E of the markings and events that occurred in the run before γ.

> **function** EXPLORE_CUT_C(γ, M, E_l)
>> $E, E_{l_{reap}} = \text{F}(\gamma, M, E_l)$
>> **if** $E = \emptyset \vee E_C(E_{l_{reap}}) \neq \emptyset$ **then return** false
>> **else**
>>> $E = \text{F}(\gamma, M)$
>>> $E_{nc} = E_{NC}(E)$
>>> $E_c = E_C(E)$
>>> **if** $E_{nc} \neq \emptyset$ **then**
>>>> $v = \text{true}$
>>>> **repeat**
>>>>> $e_0, E_{nc} = \text{EXTRACT}(E_{nc})$
>>>>> $v = \text{UNF\_EXPLORATION}(\gamma + e_0, M, e_0)$
>>>>> **if** $v == \text{true}$ **then**
>>>>>> $\text{unf} = \text{unf} \cup [\gamma, e_0, \gamma + e_0]$
>>>>> **end if**
>>>> **until** $E_{nc} == \emptyset \vee v == \text{false}$
>>>> **if** $v == true$ **then**
>>>>> $\Gamma_{good}.append(\gamma)$
>>>> **else**
>>>>> $\Gamma_{bad}.append(\gamma)$
>>>> **end if**
>>>> $ver.append(\mu(\gamma), E_{l_{reap}})$
>>>> **return** v
>>> **else**
>>>> $v = \text{false}$
>>>> **repeat**
>>>>> $e_0, E_c = \text{EXTRACT}(E_c)$
>>>>> $v = \text{UNF\_EXPLORATION}(\gamma + e_0, M, e_0)$
>>>>> **if** $v == \text{true}$ **then**
>>>>>> $\text{unf} = \text{unf} \cup [\gamma, e_0, \gamma + e_0]$
>>>>>> $\text{str} = \text{str} \cup [\gamma, e_0]$
>>>>> **end if**
>>>> **until** $E_c == \emptyset \vee v == \text{true}$
>>>> **if** $v == true$ **then**
>>>>> $\Gamma_{good}.append(\gamma)$
>>>> **else**
>>>>> $\Gamma_{bad}.append(\gamma)$
>>>> **end if**
>>>> $ver.append(\mu(\gamma), E_{l_{reap}})$
>>>> **return** v
>>> **end if**
>> **end if**
> **end function**

Algorithm 3. Events that are concurrent with the ones that already fired in the run

Input: the cut γ that must be analysed and the ordered lists M, E_l of the markings and the events that occurred in the run before γ.

Output: list of events that have been enabled from the cut associated with the first occurrence of the marking $\mu(\gamma)$ to the current cut γ, list of events that occurred in the run between the two repetitions on $\mu(\gamma)$.

    function F(γ, M, E_l)
        i = 0
        while $M[i] \neq \mu(\gamma)$ **do**
            i = i+1
        end while
        $E_{l_{reap}} = E_l[i : len(E_l)]$
        $E = []$
        for all $e \in$ ENAB_C(γ) **do**
            if $\mu(e)$ enabled in m $\forall m \in M[i : len(M)]$ **then**
                $E.append(e)$
            end if
        end for
        return $E, E_{l_{reap}}$
    end function

Algorithm 4. Full strategy

    $v = $ UNF_EXPLORATION$(\gamma_0, [], i)$
    if v == True **then**
        str = CUTS_TO_MARKINGS()
        str = COMPLETE_STRATEGY()
    end if

4. If two prefixes end with the same cut γ, it means that they differ only for the order in which the concurrent events occurred, and their possible elongations are the same. The winning condition for the User does not depend on the order in which events occurred, but only from the presence of the target in the run. Hence, if the algorithm is requested to analyse a cut for which it has already determined if α is winning, it can immediately stop and return the same answer.

5. First, we have to show that if the play does not reach the target, does not end with a deadlock, and does not reach a cut previously analysed, then this last criterion is verified. The number of reachable markings in the system is finite, hence after a number of steps equal at most to the number of reachable markings, the algorithm analyses a cut γ', such that $\mu(\gamma) = \mu(\gamma')$, where γ is a cut preceding γ' and belonging to the same play. Let us suppose that k events are enabled in γ' and concurrent with all the ones fired between γ and γ'. The algorithm adds one of these to the play and continues as before. If the play reaches a cut γ'' such that $\mu(\gamma) = \mu(\gamma'')$, then the events that the algorithm analyses are necessarily strictly less then k, because they should be concurrent

both with the events occurred between γ and γ' and with those fired between γ' and γ''. Since for every repetition, the number of events satisfying the requirements to be added decreases, after at most k cuts corresponding to the same marking $\mu(\gamma)$, the third criterion is satisfied. Notice that this does not depend on the specific cut: the same reasoning applies to all markings.

The next step is showing that if there is an $\alpha-$play with such a prefix, then there is at least an $\alpha-$play in which the User loses. We first consider the case in which there are not enabled events concurrent with all the ones in the cycle. If the prefix follows the strategy α, then the play repeating infinitely many times the behaviour of the prefix is an $\alpha-$play and the target never occurs. We cannot guarantee that the User will lose all the α-plays with such a prefix, but the fact that there is at least one is enough to state that α is not a winning strategy for the User. Secondly, we consider the case where a controllable transition fired between two occurrences of the same marking. By construction, the algorithm analyses controllable events only when all the significant uncontrollable events have been fired; hence, there cannot be any uncontrollable event that is concurrent with the cycle and that leads to the target, otherwise it would have been analysed before in the prefix. Again, if the prefix is consistent with the strategy, the play that repeats infinitely often the cycle is an $\alpha-$play and does not contain the target.

□

A consequence of Lemma 2 is the termination of the algorithm. We proved that every prefix constructed by the algorithm is finite. The number of considered $\alpha-$plays is also finite, because at every step there is only a finite number of enabled events to extend the prefix.

By construction, if the algorithm finds a winning strategy, all the runs in the prefix: (1) are consistent with the strategy, and (2) contain the target.

(1) All the plays in the list are consistent with the strategy. Every time that the algorithm analyses a cut γ and chooses to extend the prefix with a controllable event, it explores all the plays including γ and, one by one, each of the controllable enabled events. It stops when it finds a controllable enabled event such that, from the cut of the unfolding following this event, the User has a winning strategy. When this happens, the prefix is updated, adding the event and the cuts preceding and following it. Also the strategy is updated, choosing the associated controllable transition in γ. In this way, at every step, all the parts of runs in the prefix constructed until that moment are consistent with the strategy updated until that moment. If in γ there is no controllable enabled event such that, after it, the User has a winning strategy, then the part of the prefix already generated is not connected to the initial cut in the unfolding, since the event connecting this part to γ is not added to the prefix. At the same way, if there is a winning strategy, it cannot depend on the strategy calculated on the disconnected parts of the unfolding. If the algorithm finds a winning strategy and a disconnected part was found, since the algorithm chooses a controllable event in γ only when it is necessary to win, then there must be another cut in the prefix, that

precedes γ in the partial order, in which the algorithm adds a controllable transition that allows the User to avoid γ.

(2) All maximal runs in the prefix contain the target. If a run ends without the target, then the strategy allowing that run is not winning and must be changed. If it cannot be changed, then the algorithm will not state that there is a winning strategy, hence there must be a controllable node in which the decision previously taken can be changed. When another possible choice is analysed, all parts of runs depending on the previous one are deleted. Hence all the remained runs contain the target.

If the algorithm finds a winning strategy, every play in the unfolding consistent with this strategy is equivalent to an extension of a play in the prefix. This is shown in two steps.

1. Let us first consider the case without uncontrollable cyclic behaviours of the system.

   The strategy α constructed by the algorithm chooses a controllable transition only if there are not uncontrollable enabled ones. Let $\{t_1, ..., t_n\}$ be a set of uncontrollable transitions in a play, so that after their occurrence, there are not other uncontrollable enabled transitions. In whatever order the transitions are considered, the cut in the unfolding after their occurrence is the same, and the strategy will choose the same transition, because the following part of the unfolding is every time visited in the same way. Considering an α−play, there must be a prefix of its run in the unfolding, because all the uncontrollable transitions are analysed in all the uncontrollable cuts of the prefix and the strategy chooses only a transition for every cut, hence the controllable choices must be the same of the ones considered in the prefix. This is enough to state that the play is won by the User, because in the common prefix of the run there is the target transition.

2. If there are uncontrollable cyclic behaviours, such that there is a concurrent enabled transition leading to the target, then there is more variety in the possible α−plays, because the strategy is defined on markings in which uncontrollable events are enabled. Anyway, if an α-play has a prefix with the same events of one of the prefixes produced by the algorithm, then it is won by the User, regardless of the order of the cuts in the play. Some of the α−plays have a longer uncontrollable part, because if an uncontrollable transition would be finally enabled, or a controllable transition would be finally enabled and eligible, there must be a certain point in which it will fire, but the precise point is unknown. However, since we complete every cycle at least once and from every cut that is not a repetition all the possible uncontrollable extensions are explored, and since the part of the unfolding starting from a given cut is isomorphic to the part of the unfolding starting from every cut corresponding to the same marking, the uncontrollable sequence of the α−play can be divided in parts such that an isomorphic one has been considered by the algorithm.

Based on the previous observations, if the algorithm finds a winning strategy, the proposed strategy is winning in the unfolding.

Finally, we wish to show that if the algorithm states the existence of a winning strategy and proposes one on the cuts of the prefix, to complete it adding the same choice to all the markings that are part of a cycle is necessary and does not change the correctness.

- Let us suppose that a cycle with only uncontrollable transitions is in the net, and there is a controllable enabled transition that is concurrent with all the transitions in the cycle and which is necessary for the victory of the User. The strategy constructed together with the prefix adds the choice of the controllable transition only in the cut associated to the last repeated marking. This strategy is incomplete, because the chosen transition is not finally eligible, since every time that the system is in a marking of the cycle that is not the one that has been repeated in the prefix, the strategy does not choose it. To overcome this problem we fill the strategy by adding the choice of the controllable transition in every marking along the cycle.
- This preserves the correctness of the strategy. Actually, if γ is a cut that in a certain play is between γ_1 and γ_2 with $\gamma_1 < \gamma_2$ and $\mu(\gamma_1) = \mu(\gamma_2)$ and α' is the strategy computed by Algorithm 1 and translated on markings, then necessarily $\alpha'(\mu(\gamma)) = \alpha'(\mu(\gamma_1))$ or $\alpha'(\mu(\gamma)) = \emptyset$.
  Specifically, if $\alpha'(\mu(\gamma)) \neq \emptyset$, then it has to be $\alpha'(\mu(\gamma)) = \alpha'(\mu(\gamma_1))$. Let us suppose that $\{t_i\} = \alpha'(\mu(\gamma))$, $\{t_j\} = \alpha'(\mu(\gamma_2))$ and t_i precedes t_j in the enumeration defined by the input net. Then, t_i is not a winning choice in $\mu(\gamma_2)$, but there is a play that leads from γ to γ_2 and between these two cuts only uncontrollable events fire (because the controllable component is sequential). The algorithm updates the strategy in a cut only if, starting from that cut, the User is able to win every play. This cannot be the case, because the play can arrive in γ_2 and the User loses the play. Reasoning in the same way, it cannot be that t_i follows t_j in the enumeration. Hence it must be $\{t_i\} = \alpha'(\mu(\gamma)) = \alpha'(\mu(\gamma_1))$.
  If $\mu(\gamma)$ is never visited more than once in any run, then $\alpha'(\mu(\gamma)) = \emptyset$. We construct a final strategy α such that $\alpha = \alpha'$ for every marking m in which $\alpha'(m) \neq \emptyset$ and $\alpha(m') = \alpha'(m_1)$ for all m' such that there is a play in the unfolding with two cuts $\gamma_1, \gamma_2 : \mu(\gamma_1) = \mu(\gamma_2) = m_1$ and a cut $\gamma : \mu(\gamma) = m'$ and $\gamma_1 < \gamma < \gamma_2$.
  The marking m' could be reached in more than one run, and if it is part of two different uncontrollable cycles, with different repeated markings, there could be the doubt that $\alpha(m')$ is not well defined, but this is not possible. Let us suppose that there is another play in the unfolding with two cuts $\gamma_1', \gamma_2' : \mu(\gamma_1') = \mu(\gamma_2') = m_2 \neq m_1$ and a cut $\gamma' : \mu(\gamma') = \mu(\gamma) = m'$ and $\gamma_1' < \gamma' < \gamma_2'$. We have to show that if there is a winning strategy, then $\alpha'(\mu(\gamma_1)) = \alpha'(\mu(\gamma_1'))$. By contradiction, let us assume $\{t_i\} = \alpha'(\mu(\gamma_1))$, $\{t_j\} = \alpha'(\mu(\gamma_1'))$ and t_i precedes t_j in the enumeration (the opposite case is equivalent due to the symmetry of definitions). Then, t_i is not a winning choice for γ_1', otherwise it would have been chosen before analysing t_j. If t_i is not winning for γ_1', then it cannot be winning from γ_1, because, starting from γ_1 the play can arrive in γ firing only uncontrollable transitions, and from γ

there is a path made only by uncontrollable transitions to a cut γ_2'' such that $\mu(\gamma_2'') = \mu(\gamma_1')$. Since the unfolding starting from γ_2'' is isomorphic to the one starting from γ_1', if the strategy is not winning from γ_1' it cannot be winning from γ_2'' and therefore from γ_1.

4.3 Experiments

This work is mainly theoretical, and a full experimental evaluation of the algorithm is beyond the aim of this paper. However, we tested the algorithm on some preliminary examples, and we plan to extend experimentation in future works. The set of the examples that we considered and a Python implementation of the algorithm are available at https://github.com/MC3-lab/PNstrunf.

The parameters of the net that we think are important to consider are: (1) the number of elements in the net; (2) the number of controllable transitions; (3) the level of concurrency, i.e. the maximum number of concurrent transitions that are enabled in a reachable marking; (4) the presence of cycles. We evaluate the performance of the algorithm by showing the total number of calls to the functions UNF_EXPLORATION and EXPLORE_CUT_C, and the number of cuts in the prefix at the end of the execution. The results of the experiments are shown in Table 1. In all these cases, the User has a winning strategy. From the results, we see that the level of concurrency and the cycles increase the computational cost of the algorithm. In some cases, cycles raise a lot the number of necessary steps to arrive at the solutions, without contributing in the research of the strategy (this is the case in the comparison between the nets *bc* and *bc2*). We are currently working to develop a preprocessing of the net, in order to identify these inactive part that may not be considered in the research of the strategy.

Table 1. Results of the experiments

Net	$|P \cup T|$	$|K|$	Conc	Cycles	#calls	g_dim
Heart	15	2	2	no	10	8
Double_heart	26	2	3	No	31	24
Big_heart	141	30	2	No	355	126
HeartC	19	2	2	Yes	20	14
bc	23	2	3	No	19	16
bc2	27	2	4	Yes	1882	1162
10conc0	32	0	10	No	5122	1024
10conc1	32	1	10	No	2307	1025
10conc2	32	2	10	No	1028	258
conc	12	2	3	Yes	255	143

5   Other Approaches to Asynchronous Games

The general notion of asynchronous game presented in this paper was defined in [4], where it was applied to a problem of controlled liveness, under the hypothesis of full observability.

An asynchronous game on Petri nets was also defined by Finkbeiner and Olderog in [9]. This game is developed for Place/Transition nets, and is played on their unfoldings. The players are represented by tokens, moving on the places of the unfolding, divided into two teams: system and environment. System players have an equivalent function as the User in the game defined by [4] and used in this paper. Their objective is to guarantee a safety property. For example, the aim might be to avoid reaching a certain place. The places are divided into system places, where system players can move, and environment places, reserved to environment players. The strategy is defined on each place and states which is the next place where a token has to move. Places are the central elements in this game, in contrast to the game in [4] where the focus is on transitions.

The information available to the players is another difference. In [4], and in our approach, this information consists in observed transitions. If a transition is observable, then the User knows whether the transition occurred or not. If a transition is unobservable, then there is no way for the User to know whether it occurred, unless he can infer this from observations. In the game described by Finkbeiner and Olderog, the players communicate by means of synchronizations. Participating in the same transition, they acquire the knowledge of the past of the players that take part to the synchronization. One or the other approach may be more convenient for the User/System depending on the structure of the system and on the property that has to be verified.

In [9] a strategy for the System is defined on the unfolding of the net system, and must be fair, i.e. if a System player can move, then it must do it. This requirement avoids the trivial case in which safety is verified just because the players refuse to move. In the game in [4] for a similar reason, progress is granted by the environment. In that case the User wishes to force a transition to occur infinitely often. In almost every case this goal would be impossible to reach if the environment does not fire any of its transitions.

Under the restricted hypothesis of just one environment player (and an arbitrary number of system players), and complete information, Bernd Finkbeiner, Manuel Gieseking and Ernst-Rüdiger Olderog developed a tool, presented in [8], finding a strategy for the game as defined in [9]. The tool translates the game to a standard two-players game over finite graphs.

A different approach for the verification of properties through asynchronous games was developed by several authors, among which Glynn Winskel in [13] and Julian Gutierrez in [11]. The game is defined on event structures. An event structure is a set of events in which a partial order and a conflict relation are defined. Event structures are in relation with Petri nets used in this paper: given an occurrence net, there is always an event structure with the same partial order and the same conflict relations of the events in the occurrence net. The opposite is also true: constructing an occurrence net in which the partial order between events is the same as in an event structure is always possible. However, this occurrence net is not always equivalent to the unfolding of a Petri net. As in the game in [9], the two players have limited knowledge of what happens in the system. When two or more events cause the occurrence of another one, there is an exchange of information that can be used by the strategy. Gutierrez shows that the game can be applied to the bisimulation problem and model-checking.

6   Conclusions

In this work we have presented an algorithm for the computation of a strategy for a reachability problem in a distributed net system with full observability. The algorithm has been implemented and tested on different nets. The next step consists in studying its complexity.

We plan to apply the general idea of the game to different problems and to define proper algorithms to find winning strategies in each case.

On the theoretical side, we will consider the case of partial observability. In this extended case the definition of a strategy needs to be redefined, because in general, if only some transitions are observable, the current marking of the system, and the current cut on the unfolding, are unknown. Moreover, while with full observability the information given by the observations on the system or on the unfolding is the same, with partial observability a strategy on the unfolding may be able to distinguish two different evolutions of the system, even if the observed transitions are the same. This happens because in the structure of the unfolding there is a track of the different stories of the system, hence being able to distinguish two events corresponding to the same transition would mean being able to reconstruct also the unobservable story of the system up to every observed event.

In addition, we will study the possibility of implementing the strategy, by adding causal dependencies between controllable and uncontrollable transitions, formalizing them with the insertion of new places in the net. If it is possible for a winning strategy to be implemented in such a way, then the goal of the User will be reached in every execution of the obtained net system.

Another future generalization is increasing the number of players. It would be interesting to analyse a game in which more than two players try to reach a goal, eventually in a cooperative or in a competitive way, and considering a game in which concurrency is allowed also in the User component.

Acknowledgments. This work has been partially supported by MIUR. The authors thank the anonymous referees for their useful comments.

References

1. Adobbati, F., Bernardinello, L., Pomello, L.: An asynchronous game on distributed Petri nets. In: Moldt, D., Kindler, E., Wimmer, M. (eds.) Proceedings of the International Workshop on Petri Nets and Software Engineering (PNSE 2019), co-located with the 40th International Conference on Application and Theory of Petri Nets and Concurrency Petri Nets 2019 and the 19th International Conference on Application of Concurrency to System Design ACSD 2019 and the 1st IEEE International Conference on Process Mining Process Mining 2019, Aachen, Germany, June 23–28, 2019. CEUR Workshop Proceedings, vol. 2424, pp. 17–36. CEUR-WS.org (2019). http://ceur-ws.org/Vol-2424/paper2.pdf

2. Adobbati, F., Bernardinello, L., Pomello, L.: Asynchronous games on Petri nets and partial order. In: Cherubini, A., Sabadini, N., Tini, S. (eds.) Proceedings of the 20th Italian Conference on Theoretical Computer Science, ICTCS 2019, Como, Italy, September 9–11, 2019. CEUR Workshop Proceedings, vol. 2504, pp. 139–144. CEUR-WS.org (2019). http://ceur-ws.org/Vol-2504/paper17.pdf

3. Bernardinello, L., Kılınç, G., Pomello, L.: Weak observable liveness and infinite games on finite graphs. In: van der Aalst, W., Best, E. (eds.) PETRI NETS 2017. LNCS, vol. 10258, pp. 181–199. Springer, Cham (2017). https://doi.org/10.1007/978-3-319-57861-3_12

4. Bernardinello, L., Pomello, L., Puerto Aubel, A., Villa, A.: Checking weak observable liveness on unfoldings through asynchronous games. In: Moldt, D., Kindler, E., Rölke, H. (eds.) Proceedings of the International Workshop on Petri Nets and Software Engineering (PNSE2018), Bratislava, Slovakia, June 24–29, 2018. CEUR Workshop Proceedings, vol. 2138, pp. 15–34. CEUR-WS.org (2018). http://ceur-ws.org/Vol-2138/paper1.pdf

5. Best, E., Darondeau, P.: Petri net distributability. In: Clarke, E., Virbitskaite, I., Voronkov, A. (eds.) PSI 2011. LNCS, vol. 7162, pp. 1–18. Springer, Heidelberg (2012). https://doi.org/10.1007/978-3-642-29709-0_1

6. Desel, J., Kılınç, G.: Observable liveness of Petri nets. Acta Inf. **52**(2), 153–174 (2015). https://doi.org/10.1007/s00236-015-0218-1

7. Engelfriet, J.: Branching processes of Petri nets. Acta Inf. **28**(6), 575–591 (1991). https://doi.org/10.1007/BF01463946

8. Finkbeiner, B., Gieseking, M., Olderog, E.: ADAM: causality-based synthesis of distributed systems. In: Kroening, D., Păsăreanu, C.S. (eds.) CAV 2015. LNCS, vol. 9206, pp. 433–439. Springer, Cham (2015). https://doi.org/10.1007/978-3-319-21690-4_25

9. Finkbeiner, B., Olderog, E.: Petri games: synthesis of distributed systems with causal memory. Inf. Comput. **253**, 181–203 (2017). https://doi.org/10.1016/j.ic.2016.07.006

10. van Glabbeek, R.J., Goltz, U., Schicke-Uffmann, J.: On characterising distributability. Logical Methods in Comput. Sci. **9**(3) (2013). https://doi.org/10.2168/LMCS-9(3:17)2013

11. Gutierrez, J.: Concurrent logic games on partial orders. In: Beklemishev, L.D., de Queiroz, R. (eds.) WoLLIC 2011. LNCS (LNAI), vol. 6642, pp. 146–160. Springer, Heidelberg (2011). https://doi.org/10.1007/978-3-642-20920-8_17

12. Ramadge, P., Wonham, W.: The control of discrete event systems. Proc. IEEE **77**(1), 81 (1989)

13. Winskel, G.: Distributed games and strategies. arXiv preprint arXiv:1607.03760 (2016)

Solving Finite-Linear-Path CTL-Formulas Using the CEGAR Approach

Torsten Liebke$^{(\boxtimes)}$ and Karsten Wolf

Universität Rostock, Institut für Informatik, Rostock, Germany
{torsten.liebke,karsten.wolf}@uni-rostock.de

Abstract. Petri nets are an established formal method for modelling and verifying asynchronous, concurrent and distributed systems. To verify a specification, given as a temporal logic formula, state space methods often encounter the state space explosion problem. We propose a verification technique to solve the CTL query E $(\phi$ U $\psi)$ using the *Petri net state equation* with a *counterexample guided abstraction refinement* (CEGAR) approach. As a side product we show that $(EX)^k \phi$ formulas can be solved with the CEGAR approach as well. We use these special formulas as building bricks to solve the class of finite-linear-path CTL-formulas. The proposed techniques are strong at invalidating infeasible behaviour. In addition to this it will often terminate quickly. We are also introducing quick-checks for solving EG ϕ under certain circumstances.

Keywords: Petri nets · Verification · Structural analysis · CEGAR · ILP

1 Introduction

Explicit model checking algorithms encounter the state space explosion problem. A different concept to verify the reachability problem was introduced in [8] and extended by [3,4]. This concept is based on the structure of Petri nets and decreases the state space explosion problem significantly. It transforms the problem to an integer linear programming (ILP) problem, which runs iteratively based on counterexample guided abstraction refinement, proposed in [2].

Due to the fact that ILP-problems can become infeasible, the CEGAR approach is especially good to verify negative results. This makes it a valuable complement to explicit model checking algorithms, which are in general good for verifying positive results, due to the on-the-fly effect.

In [6] it is shown that it is beneficial to use specialized routines for common formulas to increase the number of verifiable problems. We propose two techniques to solve the CTL queries E$(\phi$ U $\psi)$ and $(EX)^k \phi$ with the CEGAR approach for Petri nets. Using well known tautologies, also A$(\phi$ R $\psi)$ and $(AX)^k \phi$ are solvable with these techniques. [6] also shows that only 62.3% of the E$(\phi$ U $\psi)$/A$(\phi$ R $\psi)$ formulas from the Model Checking Contest 2018 [1] are solved using the explicit model checker LoLA 2 [9]. This is due to the reason that the

© Springer-Verlag GmbH Germany, part of Springer Nature 2021
M. Koutny et al. (Eds.): ToPNoC XV, LNCS 12530, pp. 150–164, 2021.
https://doi.org/10.1007/978-3-662-63079-2_7

on-the-fly effect has no or very limited impact in some cases, e.g. when $\phi \wedge \neg\psi$ holds in the entire state space. For this case, the CEGAR approach we are introducing will terminate very quickly, stating that the ILP-problem is infeasible and thus the result of the formula is false.

We use these specialized routines as building bricks to solve a much bigger class of formulas, namely the finite-linear-path CTL-formulas. This class is characterized by two facts: First, the formulas are ending in a final marking, hence, they are finite and secondly, they have a linear witness path without branching. Using tautologies we can again check both the existential and the universal finite-linear-path formulas.

One drawback is that termination of the introduced approach is not guaranteed, which makes the procedure incomplete [8]. This drawback vanishes if a portfolio approach is applied where traditional algorithms are combined with the newly introduced methods.

We also introduce some quick-checks for verifying EG ϕ using the presence of deadlocks or the absence of certain transition invariants.

2 Basic Definitions

We consider place/transition Petri nets.

Definition 1 (place/transition net). *A* place/transition net $[P, T, F, W, m_0]$ *consists of a finite set P of places, a finite set T of transitions, a set $F \subseteq (P \times T) \cup (T \times P)$ of arcs, a mapping $W : (P \times T) \cup (T \times P) \longrightarrow \mathbb{N}$ where $[x, y] \notin F$ if and only if $W([x, y]) = 0$, and an initial marking m_0. A marking is a mapping $m : P \longrightarrow \mathbb{N}$.*

Transition t is enabled in marking m if, for all $p \in P$, $m(p) \geq W([p, t])$. Firing an enabled transition in m yields the marking m' where, for all p, $m'(p) = m(p) - W([p, t]) + W([t, p])$. This is denoted $m \xrightarrow{t} m'$.

Every Petri net defines a labeled transition system where the set of markings reachable from m_0 form the set of states, m_0 is the initial state, and the firing relation just defined forms the labeled transition relation. We restrict our considerations to Petri nets where the related transition system is finite.

The *incidence matrix* of a Petri net N is a matrix $C_N : P \times T \longrightarrow \mathbb{Z}$ where, for all $p \in P, t \in T$, $C_N(p, t) = W(t, p) - W(p, t)$. The incidence matrix is involved in important and well-known results of Petri net theory. If it is clear to which Petri net the incidence matrix belongs then we only write C.

Definition 2 (Reachability problem). *Given is a tuple (N, m, m') consisting of a Petri net N and two markings m, m'. A marking m' is reachable from marking m in a Petri net N, if there exists a firing sequence $w \in T^*$ with $m \xrightarrow{w} m'$. The set of all reachable markings in N starting in m is written as $R_N(m)$. The question whether $m' \in R_N(m)$ is called the* reachability problem.

The feasibility of the Petri net state equation is a necessary condition for a positive answer to this question.

Proposition 1 (Petri net state equation). *Let $w \in T^*$ be a firing sequence of N, that is, the sequence of labels on a path from some marking m to a marking m' in the transition system corresponding to N. Then it holds*

$$m + C \cdot \wp(w) = m'$$

where $\wp(w)$ is a vector and $|\wp(w)(t)|$ is the number of occurrences of t in the sequence w.

In the sequel, we shall refer to $\wp(w)$ as the *Parikh vector* of w.

Definition 3 (T-invariant). *A Parikh vector $\wp(w)$ is called a T-invariant if $C \cdot \wp(w) = \mathbf{0}$. If the firing sequence w is executable, we call $\wp(w)$ realizable.*

A realizable T-invariant is a cycle in the state space and will not change the marking.

Definition 4 (Solution space). *The solution of the Petri net state equation $m + C \cdot \wp(w) = m'$ can be written as the sum of a base solution and a period vector, which is a linear combination of T-invariants: $\wp(w) = b + \sum_i n_i y_i$, where $b \in \mathbb{N}^T$ is the base solution and $n_i \in \mathbb{N}$ is the coefficient of the T-invariant $y_i \in \mathbb{N}^T$ [3, 8].*

3   Increasing and Decreasing Transitions

Consider a formal sum $s = k_1 p_1 + \cdots + k_n p_n$, which we also call atomic proposition. Every marking m turns this sum into the integer number $v_s(m) = k_1 m(p_1) + \cdots + k_n m(p_n)$. We can immediately derive from the firing rule of Petri nets:

Definition 5 (Delta). *Let s be a formal sum and t a transition, then $\Delta_{t,s}$ is defined as $\Delta_{t,s} = k_1 C(p_1, t) + \cdots + k_n C(p_n, t)$.*

Lemma 1. *For all markings m, $m \xrightarrow{t} m'$ implies $v_s(m) + \Delta_{t,s} = v_s(m')$.*

Proof. Apply the Petri net state equation.   □

As we assume the transition system to be finite, there is only a finite range of values that $v_s(m)$ can take. Call an integer number k a *lower bound* for formal sum s if, for any reachable marking m, $v_s(m) \geq k$, and *upper bound* for s if, for any reachable m, $v_s(m) \leq k$. There exist several approaches in Petri net theory for computing bounds. As an example, we can solve the following optimisation problem where s is the objective function (to be minimised or maximised) and the state equation serves as side condition. If the problem yields a solution with non-diverging value for the objective function, that value is a lower (resp. upper) bound for s.

Based on Lemma 1, we can identify increasing and decreasing transitions.

Definition 6 (Increasing, decreasing). *Given an atomic proposition of the form $s \leq k$. Let L be a lower bound and U an upper bound for s. We call transition t w.r.t. the formal sum s:*

1. weakly increasing *iff* $\Delta_{t,s} < 0$
2. weakly decreasing *iff* $\Delta_{t,s} > 0$
3. strongly increasing *iff there is an upper bound U for s where $\Delta_{t,s} \leq k - U$*
4. strongly decreasing *iff there is a lower bound L for s where $\Delta_{t,s} > k - L$.*

The terminology may sound strange at first glance. However, increasing transitions have the tendency to turn a false proposition into a true one while decreasing transitions help turning a true proposition into a false one.

Let $p \leq 0$ be an atomic proposition where p is the number of tokens on place p in a Petri net. Then all transitions in the preset of p are strongly decreasing.

Lemma 2. *Consider markings m and m', transition t with $m \xrightarrow{t} m'$ and atomic proposition $s \leq k$.*

1. *If $s \leq k$ is false in m and true in m' then t is weakly increasing w.r.t. s.*
2. *If $s \leq k$ is true in m and false in m' then t is weakly decreasing w.r.t. s.*
3. *If t is strongly increasing w.r.t. $s \leq k$ then $s \leq k$ is true in m'.*
4. *If t is strongly decreasing w.r.t. $s \leq k$ then $s \leq k$ is false in m'.*

Proof. Regarding 1, we have $v_s(m) > k$ and $v_s(m') \leq k$. By Lemma 1, we conclude $\Delta_{t,s} < 0$. Regarding 3, we have $v_s(m) \geq L$ (since L is a lower bound). Hence, $v_s(m') = v_s(m) + \Delta_{t,s} \leq L + \Delta_{t,s}$ and, according to Definition 6, $v_s(m') \leq k$. □

4 CEGAR Approach for Reachability Analysis in Petri Nets

Abstraction is a powerful method for verifying systems. It omits irrelevant details of the system behaviours, to simplify the analysis and verification. Finding the right abstraction is hard. If it is too coarse, the verification might fail and if it is too fine, the state space explosion problem might occur. A solution is to use some initial abstraction [2], which is an overapproximation of the original system and then iteratively refine the abstraction based on spurious counterexamples.

In our case, the Petri net state equation is the initial abstraction for the reachability problem. Solving the state equation is a non-negative integer linear programming problem. The objective function for the ILP-problem is the shortest firing sequence of the Parikh vector $f(w) = \sum_{t \in T} |\wp(w)(t)|$ leading from the initial marking m to the final marking m'.

The feasibility of this linear system is a necessary condition for reachability, but not a sufficient one. We distinguish between three different situations:

– If the linear system is infeasible, the necessary condition is violated and the final marking is not reachable.

– If the linear system has a realizable solution, then the final marking is reachable.
– If the linear system has an unrealizable solution, which is a counterexample, then the abstraction has to be refined.

If we have an unrealizable solution, then there exists at least one $t \in T$ which fired less than $|\wp(w)(t)|$ times. To produce a new solution which avoids the spurious one, we build a refined abstraction using inequalities for the ILP-problem.

Definition 7 (Constraints). *We define two types of constraints, both being linear inequalities over transitions [8].*

– *Jump constraints have the form $|t_i| < n$, with $n \in \mathbb{N}$ and $t_i \in T$ where $|t_i|$ represents the firing count of transition t. Using the fact that base solutions are pairwise incomparable, jump constraints intend to generate a new base solution.*
– *Increment constraints have the form $\sum_{i=1}^{k} n_i|t_i| \geq n$ with $n_i \in \mathbb{Z}$, $n \in \mathbb{N}$, and $t_i \in T$. Increment constraints are used to get a new non-base solution, i.e., T-invariants are added, since their interleaving with another sequence w may turn w from unrealizable to realizable.*

Adding the two types of constrains to existing solutions we can traverse through the solution space and check whether the unrealizable solution of our linear system becomes realizable or whether the ILP-problem becomes infeasible.

Definition 8 (Partial solutions). *Let $N = (P, T, F, W, m)$ be a Petri net and $m' \in R_N(m)$ a reachability problem. A* partial solution *is a tuple $ps = (\Gamma, \wp(w), \sigma, r)$ with:*

– *Γ is the set of jump and increment constraints. Together with the state equation they form the ILP-problem.*
– *$\wp(w)$ is the minimal solution fulfilling the ILP-problem.*
– *σ is a firing sequence with $m \xrightarrow{\sigma}$ and $\wp(\sigma) \leq \wp(w)$.*
– *r is the remainder with $r = \wp(w) - \wp(\sigma)$ and $\forall t \in T : (r(t) > 0 \implies \neg m \xrightarrow{\sigma t}).$*

Partial solutions are produced during the examination of the solution $\wp(w)$ of the ILP-problem by exploring the state space of N. For this an explicit model checking algorithm with reachability preserving stubborn sets [7] can be used to build a tree of reachable markings, such that for all transitions $t \in T$ it holds that they only occur $|\wp(w)(t)|$ times. Stubborn sets, which are concerned with only one ordering of transitions, are very useful here, to avoid the explosion of the solution space. Each path to a leaf represents a maximal firing sequence of a new partial solution. If a partial solution has an empty remainder $r = 0$, it is a full solution and the reachability problem is satisfied. If no full solution exists, $\wp(w)$ might be realizable by another firing sequence σ', or by adding a jump constraint to get to a new base solution, or by adding an increment constraint to get additional tokens for transitions with $r(t) > 0$. If all possible partial solutions are explored and no full solution is found, the reachability problem can not be satisfied.

Theorem 1 (Reachability of solutions). *If the reachability problem has a solution, a realizable solution of the state equation can be reached by constantly appending the minimal solution with constraints [8].*

As stated in [3] it is an open question, whether this procedure always terminates.

5 Solving E (ϕ U ψ) with the CEGAR Approach

Definition 9 ($E(\phi$ U $\psi)$). *Let $N = (P, T, F, W, m)$ be a Petri net and ϕ and ψ two propositions. $m \models E(\phi$ U $\psi) \iff \exists w \in T^* : m \xrightarrow{w} m'$, with $\exists i \in \mathbb{N} \, \forall j < i : (m_j \models \phi) \wedge (m_i \models \psi)$. Which means that in every state along path w, ϕ is true until a state is reached where ψ is true.*

It is well known that EF ψ can be rewritten as E (true U ψ). To solve $E(\phi$ U $\psi)$, where ϕ and ψ are atomic propositions, we solve EF ψ with the CEGAR approach. In addition to this we introduce additional (balance) constraints to keep ϕ true along the path. Furthermore we cut-off paths in the exploration of partial solutions, whenever states are reached where both ϕ and ψ are false.

Definition 10 (Balance constraints). *Given a Petri net $N = (P, T, F, W, m)$ and an atomic proposition ψ and $\phi = s_0 \leq k_0 \wedge s_1 \leq k_1 \wedge \cdots \wedge s_n \leq k_n$, where s_i is a formal sum, $0 \leq i \leq n$ and $i, k, n \in \mathbb{N}$. $T_i = \{t \in T | \Delta_{t,s_i} \neq 0\}$ is the set of transitions which can change the value of s_i. It contains all weakly/strongly increasing/decreasing transitions w.r.t. to s_i. We call $T_{i,\psi} \subseteq T_i$ the set of decreasing transitions w.r.t s_i, which are at the same time increasing w.r.t ψ: $T_{i,\psi} = \{t \in T_i | \Delta_{t,s_i} > 0 \wedge \Delta_{t,\phi} < 0\}$. We define variables δ_i, which are 0, if $T_{i,\psi} = \emptyset$ and otherwise are $MAX(\Delta_{t,s_i} | t \in T_{i,\psi})$. The δ_i-offset is the maximum arc weight of all transitions that can change the value of $s_i \leq k_i$ from true to false and ψ from false to true. Let $\theta_i = k_i - v_{s_i}(m)$ be the offset, which is the number of tokens that can be consumed from the initial marking and still leave the truth value of $s_i \leq k_i$ unchanged. We call $\forall s_i : \sum_{t \in T_i} \Delta_{t,s_i} \leq \theta_i + \delta_i$ balance constraints w.r.t. s_i and m.*

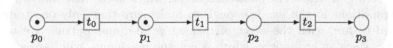

Fig. 1. The minimal solution for this Petri net and the formula E ($p_1 > 0$) U ($p_3 > 0$) is $t_1 t_2$. Since t_1 is weakly decreasing w.r.t. $p_1 > 0$, the balance constraint adds the weakly increasing transition t_0 to the solution.

As an example, consider Fig. 1 and the formula E ($p_1 > 0$) U ($p_3 > 0$). Note that this formula and every other formula can be rewritten into the required

$s \leq k$-format: E $(-p_1 \leq -1)$ U $(-p_3 \leq -1)$. To satisfy the formula, we check EF $p_3 > 0$, while keeping $p_1 > 0$ true along the path. The minimal solution to the ILP would be the firing vector (t_1, t_2), $m \xrightarrow{t_1 t_2} m'$, where m' satisfies $p_3 > 0$. But after firing the weakly decreasing transition t_1 w.r.t. $p_1 > 0$, a marking $m'' = (p_0, p_2)$ is reached that does neither satisfy $p_3 > 0$ nor $p_1 > 0$. To avoid this marking, the balance constraint would add the weakly increasing transition t_0 to the solution vector, $m \xrightarrow{t_0 t_1 t_2} m'$, to keep $p_1 > 0$ true.

Balance constraints in general ensure that the sum of all increasing and decreasing transitions w.r.t. a formal sum s is smaller than the offset, which is based on the initial marking and the maximal arc weight of all transitions $t \in T_{i,\psi}$. In case the offset θ_i is negative, ϕ is violated and E$(\phi$ U $\psi)$ has the value of ψ. We detect this case in the initial marking, before we compute the balance constraints and can return with a definitive answer directly in the beginning. Balance constraints make sure that ϕ is not violated and ψ is true in the final marking. The only transitions which are allowed to violate ϕ are in the set $T_{i,\psi}$ and they have also the effect to turn ψ to true. Due this effect, if such transitions exist, they tend to occur at the end of the firing sequence, but not exclusively. We add the balance constraints to our initial abstraction, the state equation and run the CEGAR algorithm for EF ψ.

Lemma 3. *Given a Petri net $N = (P, T, F, W, m)$ and formula $\phi = s_0 \leq k_0 \wedge s_1 \leq k_1 \wedge \cdots \wedge s_n \leq k_n$, where s_i is a formal sum and $k \in \mathbb{N}$ and $m \models \phi$. Adding to the ILP-problem all balance constraints for ϕ and checking that $\theta_i \geq 0$, then it is guaranteed that after executing the entire firing sequence given as a solution $\wp(w)$ to the ILP-problem that ψ is true. It also ensures that if a complete firing sequence exists, ϕ is true along the path and is only violated, if at all, in the final marking, where ψ holds.*

Proof. Regarding the second claim, we know, based on Definition 6, that only increasing/decreasing transitions affect $s_i \leq k_i$. The offset θ_i ensures that the truth value of $s_i \leq k_i$ stays unchanged. The balance constraint ensures that ϕ is not violated minus the δ_i-offset, which ensures the possibility of a firing sequence which does not violate ϕ along the path, until ψ holds.

If the set $T_{i,\psi}$ is not empty, the δ_i-offset based on the maximum of Δ_{t,s_i} ensures that transitions are not ignored in the balance constraint that violate ϕ but also turn ψ to true. The additional offset, which is the maximal arc weight of the transitions in the set, is enough to make sure that only one transition is allowed to fire, with the effect of making ϕ false and ψ true. We use the maximum, since an arc weight, which is not the maximum, will have a smaller effect and will not change the outcome. Transitions from the set $T_{i,\psi}$ can also fire, if they are in a different context, i.e. when they do not turn ϕ to false.

Theorem 1 ensures that if the complete solution $\wp(w)$ is fired, we get to the final marking m' which satisfies ψ. $\qquad\square$

Lemma 3 only ensures that $m' \models \psi$, where m' is the final marking after firing the entire solution $\wp(w)$. But it does not guarantee that intermediate markings satisfy ϕ. This is due to the fact that also decreasing transitions w.r.t. ϕ are allowed to fire.

Lemma 4. *In the exploration of the solution space cutting off paths in markings* m^*, *with* $m^* \models \neg\phi \wedge \neg\psi$ *results in keeping only partial solutions which can become full solutions.*

Proof. Based on Definition 9, marking $m^* \models \neg\phi \wedge \neg\psi$ violates the property $E(\phi \mathrel{U} \psi)$. All paths extending m^* are also violating $E(\phi \mathrel{U} \psi)$ and no extension to the path can make the property true. □

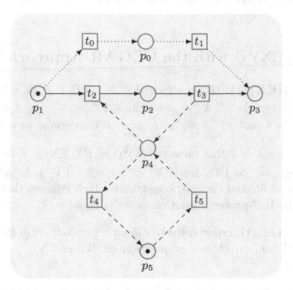

Fig. 2. For the given Petri net and the formula E $(p_1 + p_2 > 0)$ U $(p_3 > 0)$, the minimal solution (t_0, t_1) is cut off. With the CEGAR approach we jump to the next base solution (t_2, t_3), which is only a partial one. The T-invariant (t_4, t_5) is added with the next CEGAR step and provides a full solution, $m \xrightarrow{t_5 t_2 t_3 (t_4)} m'$.

Consider, for example, the Petri net in Fig. 2 and the formula E $(p_1 + p_2 > 0)$ U $(p_3 > 0)$. The minimal solution to the ILP is (t_0, t_1). After firing t_0, a marking $m' = (p_0, p_5)$ is reached that violates $p_1 + p_2 > 0$ and $p_3 > 0$. Lemma 4 ensures that this solution is cut off. There are also no increasing transitions we can add to this solution. Using the CEGAR approach, we jump to a new base solution, (t_2, t_3). But this solution is only a partial solution due to the fact that neither t_2 nor t_3 can fire. At this point, the CEGAR approach adds the T-invariant (t_4, t_5) from which tokens can be borrowed. Now we have a full solution and we get the path $m \xrightarrow{t_2 t_3 t_4 (t_5)} m'$ which satisfies $p_1 + p_2 > 0$ until $(p_3 > 0)$ is satisfied.

Theorem 2. *Let $N = (P, T, F, W, m)$ be a Petri net, ψ an atomic proposition and ϕ a proposition of the form $\phi = s_0 \leq k_0 \wedge s_1 \leq k_1 \wedge \cdots \wedge s_n \leq k_n$, where s_i is a formal sum and $i, k, n \in \mathbb{N}$ and it holds that $m \models \phi$. If $\mathrm{E}(\phi \mathrm{\ U\ } \psi)$ has a realizable solution in the solution space, it can be reached by solving EF ψ using the CEGAR approach from [8] and by adding all balance constraints to the initial abstraction and cutting-off all paths in m^* in the exploration of the solution space, whenever m^* with $m^* \models \neg\phi \wedge \neg\psi$ is reached.*

Proof. In [8] EF ψ is proved. We constantly add jump and increment constraints to get to a full solution, such that the final marking m' of this solution satisfies ψ, $m' \models \psi$. Lemma 3 ensures that we only get solutions, such that after firing the complete solution $\wp(w)$, ϕ holds. Lemma 4 makes sure that ϕ is not violated along the path.   □

6   Solving $(\text{EX})^k \phi$ with the CEGAR Approach

Definition 11 ($(\text{EX})^k \phi$). *Given a Petri net $N = (P, T, F, W, m)$, a proposition ϕ and $k \in \mathbb{N} \setminus \{0\}$. $m \models (EX)^k \phi \iff \exists w \in T^k : m \xrightarrow{w} m_k \wedge m_k \models \phi$. This means there exists a path $m \xrightarrow{w} m_k$ with $|w| = k$ transitions in it and $m_k \models \phi$.*

For example for $k = 2$ this means $(\text{EX})^2 \phi = \text{EX EX } \phi \iff \exists t_1 t_2 \in T^2 : m \xrightarrow{t_1 t_2} m_k \wedge m_k \models \phi$. To solve $(\text{EX})^k \phi$, we solve EF ϕ. In addition to this we introduce an additional (length) constraint which ensures that the length of sequence w of the ILP-problem solution $\wp(w)$ is equal to k.

Definition 12 (Length constraint). *Given a proposition of the form $(EX)^k \phi$ with $k \in \mathbb{N} \setminus \{0\}$ and an atomic proposition ϕ. We call $\sum_{t \in T} |\wp(w)(t)| = k$ a length constraint.*

The sum of the number of occurrences of all transitions in the Parikh vector $\wp(w)$ should exactly be k. To make the proposition true, marking m_k, which is reached after firing k transitions, must satisfy ϕ.

Theorem 3. *Given a Petri net $N = (P, T, F, W, m)$ and proposition $(EX)^k \phi$ with $k \in \mathbb{N} \setminus \{0\}$. If $(EX)^k \phi$ has a realizable solution in the solution space, it can be reached by solving EF ϕ using the CEGAR approach from [8] and by adding the length constraint to the initial abstraction.*

Proof. Based on Definition 11, $m \models (EX)^k \phi \iff \exists w \in T^k \wedge m \xrightarrow{w} m' \wedge m' \models \phi$. The length constraint $\sum_{t \in T} |\wp(w)(t)| = k$ from Definition 12 ensures that only solutions $\wp(w)$ of the ILP-problem are found, such that the length of the firing sequence is exactly k and results in the final marking $m_k \models \phi$.   □

7 Solving Finite-linear-path CTL-formulas with the CEGAR Approach

Theorems 1-3 are solving simple CTL-formulas. They all have in common that they have a linear and finite witness path. We use this building bricks to solve a larger class of CTL-formulas with the CEGAR approach. Namely the class of finite-linear-path formulas.

Definition 13 (Existential finite-linear-path formula). *If ϕ and ψ are existential finite-linear-path formulas and ρ is an atomic proposition, then the following formulas are existential finite-linear-path formulas:*

- *ρ (the base of the inductive definition);*
- *EF ϕ;*
- *EX ϕ;*
- *E(ρ U ϕ);*
- *$\phi \vee \psi$;*
- *$\phi \wedge \rho$;*

The existentially quantified formulas are paired with the universally quantified formulas. These two formulas can be reduced to each other by negation. Hence, they permit the application of the same verification techniques. The class of CTL-formulas is extended to the universal finite-linear-path formulas, which use the path as a counterexample. The class is defined accordingly:

Definition 14 (Universal finite-linear-path formula). *If ϕ and ψ are universal finite-linear-path formulas and ρ is an atomic proposition, then the following formulas are universal finite-linear-path formulas:*

- *ρ (the base of the inductive definition);*
- *AG ϕ;*
- *AX ϕ;*
- *A(ρ R ϕ);*
- *$\phi \wedge \psi$;*
- *$\phi \vee \rho$;*

It is easy to see that the negation of an existential finite-linear-path formula is indeed a universal finite-linear-path formula and vice versa. That is, we may restrict subsequent considerations to existential finite-linear-path formulas. Universal finite-linear-path formulas can be verified by checking their negation.

We introduce the concept of how to solve this class of formulas with an example. The interesting part of this class are formulas which have nested CTL-operators, e.g. E (ρ_1 U (E (ρ_2 U ϕ))). The idea is to use for each CTL-operator one state equation with its own set of variables and constraints and then solve the entire ILP-problem.

In our example the first objective would be to solve the left/outer EU-formula. That is, we have to reach a marking $m' \models \rho_2$ while keeping ρ_1 true. For this we have to solve the ILP-problem consisting of the state equation

$m + C \cdot \wp(w)_1 = m'$ and the balance constraints for ρ_1. The second objective is to solve the right/inner EU-formula. Here we are doing the same things as before, that is, we have to reach a marking $m'' \models \phi$ while keeping ρ_2 true. We now add to the ILP-problem a slightly different state equation, $m' + C \cdot \wp(w)_2 = m''$, where we start in the marking m', which we reached from the first state equation and furthermore we introduce a new set of variables $\wp(w)_2$ for our second Parikh vector to reach the final marking m''. The balance constraints to keep ρ_2 true are added as well. Both state equations can be linked together into one equation, $m + C \cdot \wp(w)_1 + C \cdot \wp(w)_2 = m''$.

Definition 15 (ILP-problem for existential finite-linear-path formula).
Let $N = (P, T, F, W, m)$ be a Petri net and ϕ be an existential finite-linear-path formula, which contains $i \in \mathbb{N}$ CTL-operators. We call the following an ILP-problem for an existential finite-linear-path formula or in short ILP_ϕ:

For all CTL-operators add a new set of variables for the Parikh vector $\wp(w)_i$ and the product of $C \cdot \wp(w)_i$ to the state equation:

$$m + C \cdot \wp(w)_1 + \ldots + C \cdot \wp(w)_i = m'.$$

Also add for all EU-operators balance constraints and for EX-operators length constraints based on their corresponding variables.

Once we build the initial ILP-problem we can use the CEGAR approach to find either a realizable solution or to add enough constraints to make the ILP-problem infeasible to verify that no solution exists. While realizing the solution it is important to first use all the transitions from the first Parikh vector $\wp(w)_1$ to keep the structure of the formula in place. $\wp(w)_1$ keeps ρ_1 true until ρ_2 is reached. If all transitions from $\wp(w)_1$ are used in the realization we can start with the transitions of $\wp(w)_2$.

Definition 16 (Realization ordering). *Let $N = (P, T, F, W, m)$ be a Petri net, ϕ an existential finite-linear-path formula, which contains $i \in \mathbb{N}$ CTL-operators and ILP_ϕ the corresponding ILP-problem. To keep the structure of ϕ in place while realizing a solution of ILP_ϕ it must hold that $\forall j, k \in \mathbb{N} : 0 \leq j < k \leq i$ the transitions from $\wp(w)_j$ must be realized before the transitions of $\wp(w)_k$. We call this the* realization ordering.

Theorem 4. *Let $N = (P, T, F, W, m)$ be a Petri net, ϕ be an existential finite-linear-path formula and ILP_ϕ be an ILP-problem for the existential finite-linear-path formula ϕ based on Definition 15. If ϕ has a realizable solution in the solution space, it can be reached by using Theorems 1–3 with ILP_ϕ as the initial ILP-problem and using the realization ordering based on Definition 16 for finding a realizable solution.*

Proof. We proceed by induction, according to Definition 13.
Case ρ (atomic proposition): In CTL an atomic proposition is satisfied, if it holds in the initial marking. Based on Definition 15 and the fact that no CTL-operator is present, no product of $C \cdot \wp(w)$ is added to the equation. It follows

that $m = m'$, which means that the atomic proposition must hold in the initial marking.

Case EF ϕ: This case can be traced back to Case $E(\rho \cup \phi)$ using the tautology EF $\phi \iff$ E(TRUE U ϕ).

Case EX ϕ: Definition 15 ensures that $C \cdot \wp(w)$ is added to the state equation and that the length constraint for EX ϕ is added to the ILP-problem. A witness path for EX ϕ is an existential finite-linear-path to the next marking which satisfies ϕ. The path extended by a witness path for ϕ at the final marking (which exists by induction hypothesis) yields a witness path for EX ϕ. Theorem 3 makes sure that if a realizable solution exists, the witness path for EX ϕ is found and Definition 16 ensures that the witness path is added at the correct position to keep the structure of the formula in place.

Case $E(\rho \cup \phi)$: This case is similar to the previous one. Definition 15 ensures that $C \cdot \wp(w)$ is added to the state equation and that the balance constraints are added to the ILP-problem. A witness path for $E(\rho \cup \phi)$ is an existential finite-linear-path where ρ is true in every marking until a marking is reached where ϕ holds. Theorem 2 makes sure that if a realizable solution exists, the witness path for $E(\rho \cup \phi)$ is found and Definition 16 ensures that the witness path is added at the correct marking (which exists by induction hypothesis) to keep the structure of the formula in place.

Case $\phi \vee \psi$: If ϕ is satisfied then there exists a witness path for ϕ for which the induction hypothesis may be applied. Otherwise, there is a witness path for ψ for which again the induction hypothesis applies. A formula like EX $\phi \vee \psi$ is rewritten to EX $\phi \vee$ EX ψ and both sides are verified separately.

Case $\phi \wedge \rho$: In this case, ϕ and ρ are satisfied. Since ρ is an atomic proposition, only the initial marking of the path is concerned. Hence, the induction hypothesis applied to ϕ yields the desired result.   □

8   Partially Solving EG ϕ with the CEGAR Approach

Definition 17 (EG ϕ). *Let $N = (P, T, F, W, m)$ be a Petri net and ϕ a proposition. $m \models$ EG $\phi \iff \exists w \in T^* : m \xrightarrow{w} m'$, with $\forall i \in \mathbb{N} : (m_i \models \phi)$. This means that in every state along a path w, ϕ is true.*

Definition 18 (DEADLOCK). *Given a Petri net $N = (P, T, F, W, m)$. N has a deadlock if there exist a reachable marking from m in which no transition is activated.*

ϕ is true along a path w, if at least one of two conditions is fulfilled. Either there exists an infinite path containing a cycle or the path ends in a deadlock. Precisely:

1. If the path is infinite then there exists a cycle and the path can be split into two parts $w_1 w_2$ with $m \xrightarrow{w_1} m' \xrightarrow{w_2} m'$, where w_1 is a path leading to a marking m', from which a cycle starts, namely w_2, which goes back to m'. Each state in both w_1 and w_2 satisfies ϕ and w_2 can be repeated infinitely often.

2. If the paths ends in a deadlock every state including the last one, the deadlock state, must satisfy ϕ.

In both cases we can use the knowledge about the existence of a deadlock to create necessary or sufficient quick-checks to solve EG ϕ. If the Petri net has no deadlocks, the only possibility to satisfy EG ϕ is if a cycle can be reached while ϕ stays true and the cycle keeps ϕ also true in every state. The cycle is basically a T-invariant and we can reformulate the problem of solving EG ϕ into solving the state equation once and finding a T-invariant while keeping ϕ true,

$$m \xrightarrow{m + C \cdot \wp(w_1) = m'} m' \xrightarrow{C \cdot \wp(w_2) = 0} m'.$$

Solving the first part, the state equation, is problematic due to fact that m' is not known. The reason for this is that there can be exponentially many T-invariants which keep ϕ true in every state. In addition to this we would have to solve the problem of finding a minimal marking to fire a T-invariant, where minimal is in regard to the entire token number in the marking. It would also make no difference if minimal is in regard to the componentwise comparison of markings, meaning that no more token can be removed. To the best of our knowledge there is no polynomial algorithm known for this problem. We could use a brute-force-method where we calculate for every sequence of a T-invariant, which are all permutations, the minimal required markings to fire completely. All markings can then be compared and we can search for the minimal markings. The runtime for this method would be exponential. This, in connection with the possibility of exponentially many T-invariants, is not a suitable approach to solve EG ϕ.

But on the other hand the second part can be used to build a necessary condition check. If no T-invariant exists that keeps ϕ true and the Petri net has no deadlocks we know that EG ϕ can never be true. To check this we can add to the ILP-problem for finding an invariant an adjusted version of the balance constraint from Definition 10.

Definition 19 (Minimum constraints). *Let $N = (P, T, F, W, m)$ be a Petri net and a proposition $\phi = s_0 \leq k_0 \wedge s_1 \leq k_1 \wedge \cdots \wedge s_n \leq k_n$, where s_i is a formal sum, $0 \leq i \leq n$ and $i, k, n \in \mathbb{N}$. $T_i = \{t \in T | \Delta_{t,s_i} \neq 0\}$ is the set of transitions which can change the value of s_i. It contains all weakly/strongly increasing/decreasing transitions w.r.t. to s_i. We call $\forall s_i : \sum_{t \in T_i} \Delta_{t,s_i} \leq 0$ minimum constraints w.r.t. s_i.*

These constraints ensure that the sum of all increasing and decreasing transitions is smaller than or equal to zero. Otherwise the truth value of the proposition will be changed.

Proposition 2. *Given a Petri net $N = (P, T, F, W, m)$ and a proposition $\phi = s_0 \leq k_0 \wedge s_1 \leq k_1 \wedge \cdots \wedge s_n \leq k_n$, where s_i is a formal sum and $i, k, n \in \mathbb{N}$ and it holds that $m \models \phi$. If the Petri net has no deadlocks and if the ILP-problem for finding a T-invariant, $C \cdot \wp(w) = \mathbf{0}$ in addition with the minimum constraints has no solution, then EG ϕ is also false, $m \not\models EG \phi$.*

Proof. Based on Definition 17 if the Petri net has no deadlocks then the only way to satisfy EG ϕ is to find a cycle which keeps ϕ true in every state. If there exists such a cycle it must be a T-invariant and based on Definition 3 the equation $C \cdot \wp(w) = \mathbf{0}$ must have a solution. The minimum constraints based on Definition 19 ensure that ϕ stays true in the cycle. If the ILP-problem, $C \cdot \wp(w) = \mathbf{0}$ plus the minimum constraints, is infeasible, then no T-invariant, therefore no cycle exists, that keeps ϕ true. It follows that EG ϕ can never be true. □

In case the Petri net has deadlocks we can build a sufficient quick-check. We use the fact that EG ϕ is true if the path ends in a deadlock and every state along the path satisfies ϕ. In CTL this condition can be rewritten to $E(\phi \text{ U } (\phi \land \text{DEADLOCK}))$, where the DEADLOCK predicate can be easily expressed as a conjunction of disjunctions over atomic propositions.

Proposition 3. *Given a Petri net $N = (P, T, F, W, m)$ with deadlocks and an atomic proposition ϕ. If the ILP-problem for $E(\phi \text{ U } (\phi \land \text{DEADLOCK}))$ has a realizable solution, then EG ϕ is true, $m \models$ EG ϕ.*

Proof. If the Petri net has deadlocks, then based on Definition 17 EG ϕ is among others true, if a path which satisfies ϕ in every state ends in a deadlock. Definition 9 states that ϕ is true until ψ holds and ψ is in this case $\phi \land \text{DEADLOCK}$. □

9 Conclusion and Future Work

We proposed two promising techniques to solve $E(\phi \text{ U } \psi)$ and $(\text{EX})^k \phi$ with the CEGAR approach for Petri nets and used this as building bricks to solve the class of finite-linear-path CTL-formulas. The main concept is to use constraints on the Parikh vector. We refine the over approximation iteratively until it becomes a realizable solution or infeasible. We also introduced quick-checks for solving EG ϕ under certain circumstances.

To solve $E(\phi \text{ U } \psi)$, we solve EF ψ and keep ϕ true in every state along the path. To keep ϕ true, we introduced the concept of balance constraints for the ILP-problem to ensure that an atomic proposition is true after firing the entire solution vector. Furthermore we used a cut-off criterion to ensure that ϕ is also true in every state along the path. For solving $(\text{EX})^k \phi$ we introduced the concept of a length constraint, which makes sure that we only get solutions of length k. The finite-linear-path formulas are using the proposed techniques for solving $E(\phi \text{ U } \psi)$ and $(\text{EX})^k \phi$ in addition to an ILP-problem that is build dependent on the CTL-operators contained in the finite-linear-path formula. To verify EG ϕ with a necessary quick-check in the absence of deadlocks we proposed a minimum constraints which ensure that when no T-invariant is found, EG ϕ must be false. As a sufficient quick-check in the presence of deadlocks we introduced the deadlock-constraint and check if $E(\phi \text{ U } (\phi \land \text{DEADLOCK}))$ has a realizable solution. All proposed techniques are based on solving ILP-problems and thus avoiding the state space explosion problem.

These techniques will be implemented in LoLA 2 [9]. LoLA 2 is an explicit model checker and is every year on the podium of the Model Checking Contest for Petri nets. Once implemented we expect that the proposed approach will increase the verification performance for this formulas significantly. Especially in case of a negative result, the procedure will terminate quickly, due to the fact that the ILP-problem will become infeasible. We expect a similar performance increase as it was the case for the CEGAR approach for reachability analysis, where the performance of LoLA 2 increased from solving under 80% to over 90% in the Model Checking Contest.

Acknowledgements. This study is an extended version of [5]. We thank the anonymous reviewers of both PNSE and ToPNoC for their comments.

References

1. Amparore, E.G., et al.: Presentation of the 9th edition of the model checking contest. In: Tools and Algorithms for the Construction and Analysis of Systems - 25 Years of TACAS: TOOLympics, Held as Part of ETAPS 2019, 6–11 April 2019, Prague, Czech Republic, Proceedings, Part III, pp. 50–68 (2019)
2. Clarke, E.M., Grumberg, O., Jha, S., Lu, Y., Veith, H.: Counterexample-guided abstraction refinement. In: Computer Aided Verification, 12th International Conference, CAV 2000, 15–19 July 2000, Chicago, IL, USA, Proceedings, pp. 154–169 (2000)
3. Hajdu, Á., Vörös, A., Bartha, T.: New Search strategies for the petri net CEGAR approach. In: Devillers, R., Valmari, A. (eds.) PETRI NETS 2015. LNCS, vol. 9115, pp. 309–328. Springer, Cham (2015). https://doi.org/10.1007/978-3-319-19488-2_16
4. Hajdu, Á., Vörös, A., Bartha, T., Mártonka, Z.: Extensions to the CEGAR approach on Petri nets. Acta Cybern. **21**(3), 401–417 (2014)
5. Liebke, T., Wolf, K.: Solving E (ϕ U ψ) using the CEGAR approach. In: Moldt, D., Kindler, E., Wimmer, M. (eds.) Petri Nets and Software Engineering. International Workshop, PNSE 2019, Aachen, Germany, June 24, 2019. CEUR Workshop Proceedings. CEUR-WS.org, vol. 2424, pp. 47–56 (2019)
6. Liebke, T., Wolf, K.: Taking some burden off an explicit CTL model checker. In: Donatelli, S., Haar, S. (eds.) PETRI NETS 2019. LNCS, vol. 11522, pp. 321–341. Springer, Cham (2019). https://doi.org/10.1007/978-3-030-21571-2_18
7. Schmidt, K.: Stubborn sets for standard properties. In: Application and Theory of Petri Nets 1999, 20th International Conference, ICATPN 1999, 21–25 June 1999, Williamsburg, Virginia, USA, Proceedings, pp. 46–65 (1999)
8. Wimmel , H., Wolf, K.: Applying CEGAR to the Petri net state equation. In: Tools and Algorithms for the Construction and Analysis of Systems - 17th International Conference, TACAS 2011, Held as Part of the Joint European Conferences on Theory and Practice of Software, ETAPS 2011, 26 March – 3 April, Saarbrücken, Germany, 2011. Proceedings, pp. 224–238 (2011)
9. Wolf, K.: Petri net model checking with LoLA 2. In: Application and Theory of Petri Nets and Concurrency - 39th International Conference, PETRI NETS 2018, 24–29 June 2018, Bratislava, Slovakia, Proceedings, pp. 351–362 (2018)

Verification of the MQTT IoT Protocol Using Property-Specific CTL Sweep-Line Algorithms

Alejandro Rodríguez[(✉)], Lars Michael Kristensen, and Adrian Rutle

Department of Computer Science, Electrical Engineering, and Mathematical Sciences,
Western Norway University of Applied Sciences, Bergen, Norway
{arte,lmkr,aru}@hvl.no

Abstract. MQTT is a publish-subscribe communication protocol being increasingly used for implementing internet-of-things (IoT) applications. In earlier work we have developed a formal and executable model of the MQTT protocol using Coloured Petri Nets (CPNs) and performed an initial verification of behavioural properties. The contribution of this paper is to investigate the use of the sweep-line method for verification of the MQTT CPN model in order to alleviate the effect of the state explosion problem. We formulate the behavioural properties using Computation Tree Logic (CTL) and show how to formulate a progress measure for the sweep-line method based on the main phases of the MQTT protocol. To perform the verification of properties, we provide some property-specific CTL model checking algorithms compatible with the sweep-line method.

Keywords: Coloured Petri Nets · Modelling · Verification · Communication protocols · Internet of Things

1 Introduction

The development of distributed software systems is challenging, and one of the main approaches to tackle the challenges is to build an executable model of the system prior to implementation and deployment. Coloured Petri Nets (CPNs) [13] is a formal modelling formalism convenient for specifying complex concurrent and distributed systems. CPN Tools [9,15] is a software tool that supports the construction, simulation (execution), state space analysis, and performance analysis of CPN models. One of the key functionalities of CPN Tools is the ability to perform model checking [1] of the modelled system. This means that one can generate the state space (the set of reachable states) of a system in order to verify key behavioural properties. Temporal logics [23] such as Computation Tree Logic (CTL) and Linear Temporal Logic (LTL) are widely used to express behavioural properties of systems.

MQTT [2] is a publish-subscribe messaging protocol for IoT suited for constrained application domains such as Machine-to-Machine communication

© Springer-Verlag GmbH Germany, part of Springer Nature 2021
M. Koutny et al. (Eds.): ToPNoC XV, LNCS 12530, pp. 165–183, 2021.
https://doi.org/10.1007/978-3-662-63079-2_8

(M2M) and IoT contexts. MQTT is designed with the aim of being light-weight and easy to implement. In earlier work [19], we have developed a formal and executable specification of MQTT motivated by the fact that until now, the protocol has only been specified using an (ambiguous) natural language specification. MQTT contains relatively complex protocol logic for handling connections, subscriptions, and quality of service levels related to message delivery.

Our initial verification experiments were conducted using ordinary full state spaces and clearly highlighted the presence of the state explosion problem [8,22]. This was caused by the exponential growth in the number of reachable states of the system with respect to the number of clients, packets, and topics. A large part of the model checking research has aimed at developing techniques for alleviating this inherent complexity problem. This includes several different families of reduction methods such as partial-order reduction methods [7] that reduce the number of interleaving execution considered, and hash compaction [21] which provides a compact representation of states with a small probability of not covering the complete state space. Since the amount of memory is often the limiting factor in model checking, we focus on the family of methods that combat state explosion by deleting states from memory during state space exploration. Specifically, we consider the sweep-line method [12] which is based on the idea of exploiting a notion of progress exhibited by many systems. We focus on CTL because CPN Tools implements a CTL-based temporal logic called ASK-CTL [3] which enables queries taking into account both state and event information. Furthermore, CTL is able to capture the behavioural properties of interest for the MQTT protocol.

The contribution of this paper is twofold: (1) the implementation of the sweep-line method using the Standard ML (SML) language together with the ability of performing model checking of certain behavioural properties specified using tailored CTL sweep-line model checking algorithms based on [17]; and (2) the application of sweep-line based CTL model checking to our CPN model of the MQTT IoT protocol. It should be noted that there already exists work on LTL model checking using the sweep-line method [10], but several of the behavioural properties that we aim to verify for MQTT are true CTL properties, i.e., not expressible in LTL [22,24].

The rest of this paper is organised as follows. In Sect. 2 we introduce the sweep-line method and in Sect. 3 we provide the property-specific CTL model checking algorithm that we employ for the verification. Section 4 gives a brief review of the CPN model of the MQTT protocol. We describe the experiments carried out and the results obtained in Sect. 5. Finally, in Sect. 6, we sum up the conclusions and outline directions for future work. The reader is assumed to be familiar with the basic concepts of CPNs and CTL model checking techniques. This paper is based upon the workshop paper [20] and the conference paper [17].

2   The Sweep-Line State Space Exploration Method

The sweep-line method [4] is aimed at systems for which it is possible to define a measure of progress based on the states of the system. A progress measure maps each state of the system into a *progress value* and is in most cases specific for the system under consideration. In this paper, we consider the version of the sweep-line algorithm for *monotonic progress measures*. The key property of a monotonic progress measure is that for any given state s, all states reachable from s have a progress value which is greater than or equal to the progress value of s. This means that a monotonic progress measure preserves the reachability relation. Having defined a progress measure of the system makes it possible to organise the state space into *layers* such that states that share the same progress value belong to the same layer.

The basic idea of the sweep-line method is to explore the state space in a least-progress-first order, one layer at a time, such that once all states in a given layer have been processed, they are removed from memory and the exploration proceeds to the next layer [12]. In conventional state space exploration, the states are kept in memory to recognise already visited states. However, a monotonic progress measure guarantees that states which have a progress value that is strictly less than the minimal progress value of those states for which successors have not yet been calculated can never be reached again. It is therefore safe to delete such states from memory which significantly reduces the memory usage during the state space exploration.

The progress exploited by the sweep-line method and formalised in the form of a progress measure is defined below in Definition 1 where S denotes the set of system states, $s_0 \in S$ denote the initial state, $s \to^* s'$ denotes that $s' \in S$ is reachable from $s \in S$ via some number of transitions, and $reach(s_0)$ the set of states reachable from the initial state.

Definition 1 (Monotonic Progress Measure). A **monotonic progress measure** is a tuple $\mathcal{P} = (O, \sqsubseteq, \Psi)$ such that O is a set of **progress values**, $\sqsubseteq$ is a total order on O, and $\Psi : S \to O$ is a **progress mapping** such that $\forall s, s' \in reach(s_0) : s \to^* s' \Rightarrow \Psi(s) \sqsubseteq \Psi(s')$. $\qquad\qquad\square$

A progress measure is non-monotonic when there is at least one *regress edge*, i.e., an edge where the source state has a larger progress value than the destination state. A generalised version of the sweep-line method that can handle non-monotonic progress measures and regress edges also exists [14], but is not the focus of our work. It was already proved [12] that the sweep-line method guarantees full coverage of the state space, and in the case of a monotonic progress measure it terminates after having explored each reachable state once. In the case of a non-monotonic progress measures, termination is still guaranteed but some states may be explored multiple times.

Algorithm 1 based on [12] specifies the sweep-line algorithm for monotonic progress measures. The algorithm starts with a hash table of visited states and a priority queue on progress values containing the states that are still to be processed. Both are initialized at the beginning with the initial state s_0 (lines 2-3). The progress value for the current (initial) layer ψ_c is also initialized in line 4. Then, the algorithm executes a loop (lines 5-28) which ends when all the reachable states have been processed. For each iteration, we select one of the

```
    Data:
    Nodes ▷ Hash table of visited states currently stored.
    Unprocessed ▷ Priority queue of unprocessed states.
    Layer ▷ List of states processed in the current layer.
    ψc ▷ Progress value for current layer.
    Φ ▷ Property to be verified.
    Result: True if the property is satisfied, false otherwise.
 1  begin
 2  │   Nodes.insert(s0)
 3  │   Unprocessed.insert(s0)
 4  │   ψc ⟵ ψ(s0)
 5  │   while ¬(Unprocessed.isEmpty()) do
    │   │   /* node with lowest progress measure                    */
 6  │   │   s ⟵ Unprocessed.getMinElement()
 7  │   │   if ψc ⊏ ψ(s) then
 8  │   │   │   if ¬ (checkProperty(Layer, Φ)) then
 9  │   │   │   │   return false
10  │   │   │   end
11  │   │   │   forall s′ ∈ Layer do
12  │   │   │   │   Nodes.delete(s′)
13  │   │   │   end
14  │   │   │   Layer ⟵ ∅
    │   │   │   /* Update progress measure for current layer        */
15  │   │   │   ψc ⟵ ψ(s)
16  │   │   end
17  │   │   Layer.insert(s)
    │   │   /* For every successor state of s                       */
18  │   │   forall (t, s′) such that s ⟶ᵗ s′ do
19  │   │   │   if ¬(Nodes.contains(s′)) then
20  │   │   │   │   Nodes.insert(s′)
21  │   │   │   │   if (ψ(s) ⊐ ψ(s′)) then
22  │   │   │   │   │   RaiseException('Regress edge found')
23  │   │   │   │   else
24  │   │   │   │   │   Unprocessed.insert(s′)
25  │   │   │   │   end
26  │   │   │   end
27  │   │   end
28  │   end
29  │   return true
30  end
```

Algorithm 1: Sweep-line algorithm for monotonic progress measures

states with the lowest progress value among the unprocessed states (line 6). The condition in line 7 checks if the progress value of the layer is strictly less than the progress value of the selected state; if so, we are about to move into the next layer. This is the point where we invoke the property-specific CTL model checking algorithm for the property Φ using the CHECKPROPERTY procedure at line 8. If the CHECKPROPERTY determines that the property is violated, then we return false and the algorithm stops. The implementation of CHECKPROPERTY is the subject of the next section. In line 18, we use $s \xrightarrow{t} s'$ to denote that the transition t is enabled in state s, and that the occurrence of t in s leads to the state s'. If the property is never violated the algorithm returns true at the end of the execution (line 29).

3   CTL Property Checking Algorithms

CTL [5] is an important branching temporal logic that is sufficiently expressive for the formulation of an important set of behavioural system properties. Even though a large set of properties can be specified using the semantics of CTL, there are some restrictions when applying them with the sweep-line method algorithm. The challenge of combining CTL model checking with the sweep-line method is that conventional algorithms for CTL model checking propagate information backwards from a state to its predecessors [6]. This follows the opposite workflow than the forward progress-first exploration that the sweep-line method performs.

In this paper, we do not consider the full CTL, but only formulas of the $AG\{EF, AF\}$-fragment that can be obtained from the following grammar, where p as an atomic state proposition and ϕ is called a *state predicate*:

$$\Phi ::= \mathbf{AG}\,\psi \mid \psi$$
$$\psi ::= \mathbf{EF}\,\phi \mid \mathbf{AF}\,\phi \mid \phi$$
$$\phi ::= p \mid \phi_1 \wedge \phi_2 \mid \phi_1 \vee \phi_2 \mid \neg\phi$$

The formulas expressing behavioural properties to be verified are interpreted over the paths of the state space as informally explained below:

Property - AG ψ "Invariantly", which holds if ψ holds in all states that are reachable from the current state.

Property - EF ϕ "Holds potentially" or "possibly", which holds if it is possible to find a state reachable from the current state where ϕ holds.

Property - AF ϕ "Holds eventually" which holds if from the current state, a state satisfying ϕ is always eventually reached.

Property - AG EF ϕ "Always possible", which holds if from any state reachable from the current state, a state satisfying ϕ can always be reached.

Property - AG AF ϕ "Always eventually", which holds if from any state reachable from the current state, a state satisfying ϕ is always eventually reached.

We say that a formula (property) Φ holds if Φ holds in the initial state s_0. To model check the **AF EF** and **AG AF** properties, we exploit the set of *strongly connected components (SCC)*. A strongly connected component of a directed graph is a maximal subgraph determined by nodes that are mutually reachable. A strongly connected component is *terminal* if no states in the component has outgoing edges to states in other components. It should be noted that when checking the **AG AF** and **AF** properties we implicitly add a self-loop to any terminal states, i.e. (deadlocked) states without enabled transitions.

Because of the monotonicity of the progress measure, each strongly connected component only contains nodes belonging to the same layer and is hence always contained in a single layer. This is formally stated in the proposition below.

Proposition 1. *Let* $\mathcal{P} = (O, \sqsubseteq, \psi)$ *a monotonic progress measure, SCC be the set of strongly connected components, and let scc $\in SCC$ be a strongly connected component. Then:* $\forall s, s' \in scc : \psi(s) = \psi(s')$.

Proof. Assume that there exists an $scc \in SCC$ and states $s, s' \in scc$ such that $\psi(s) \neq \psi(s')$. Hence either $\psi(s) \not\sqsubseteq \psi(s')$ or $\psi(s') \not\sqsubseteq \psi(s)$. Since s and s' are in the same scc, then they are mutually reachable and therefore there must exist a pair of states (s_i, s_j) on the path from either s to s' or s' to s such that $\psi(s_i) \not\sqsubseteq \psi(s_j)$. This contradicts the fact that the progress measure is monotonic.

Based on this, we can compute the strongly connected components for a given layer immediately before we delete the nodes in the current layer and move to the next one. The algorithm checks the property depending on the form of the property as outlined below.

Property - AG ϕ. We check that every node within the layer satisfies ϕ. If ϕ does not hold in one of them, we return false and abort the exploration.

Property - EF ϕ. If at least one state is encountered that satisfies ϕ, then true is returned and the execution finishes. Thus, false will be returned if at the end of the exploration not a single state satisfying ϕ has been found.

Property - AG EF ϕ. For this property, we first compute the SCC of the given Layer. The property will not be satisfied and therefore the procedure will finish the execution returning false, if any scc among the SCC of Layer is terminal and ϕ does not hold in any of the states contained in scc.

Property - AG AF ϕ. For this property, we first compute the SCC of the given Layer. We then remove the states that satisfy ϕ. If the resulting set of nodes has a cycle, then the property is violated and therefore the execution immediately finishes returning false.

Property - AF ϕ. This property can be checked in a similar fashion as **AG AF** ϕ with the modification that we can truncate the search at SCC where all cycles include a state satisfying ϕ.

The two first properties can easily be checked by just inspecting each state encountered during the sweep-line state space exploration. For verification of the two other properties, we invoke the procedure CHECKPROPERTY at the moment where the algorithm is about the leave the current layer and move into the next ones. We do not detail the checking of **AF** ϕ as it is very similar to **AG AF** ϕ as explained above.

A consequence of Proposition 1 is that SCC can be computed by considering one layer at a time. Furthermore, Theorem 1 ensures that the sweep-line method covers all reachable states which means that we will encounter all strongly connected components at some stage. The remaining step consist of linking the inspection of SCC to the model checking of the **AG EF** and **AG AF** properties. This is done in the proposition below which formalises the requirements informally introduced above.

Proposition 2. *Let SCC be the set of strongly connected components of M, $SCC_T \subseteq SCC$ the set of terminal strongly connected components, and let ϕ be a state predicate. Then:*

1. **AG EF** ϕ *is satisfied* $\Leftrightarrow \forall scc \in SCC_T \; \exists s \in scc : \phi(s)$
2. **AG AF** ϕ *is satisfied* $\Leftrightarrow \forall scc \in SCC : scc \setminus \{s \in scc : \phi(s)\}$ *is acyclic*

Proof. First we prove 1. Assume that **AG EF** ϕ holds and there exists a terminal scc named scc_t such that no states in scc_t satisfy ϕ. Since all states belong to some scc, then we can find a path from the initial state to a state s in scc_t. Since scc_t is terminal and do not contain states satisfying ϕ, then we can no longer reach states that satisfies ϕ from s. Hence, **AG EF** ϕ cannot hold. Assume that each terminal scc contains a state satisfying ϕ and let s be any reachable state. Since we cannot have cycles that spans multiple SCC and all states belong to some scc, there must exists a path from the scc to which s belongs to a state s' in some terminal scc. Within this terminal scc, all states are mutually reachable and by our assumption at least one state in there satisfies ϕ. Hence, **AG EF** ϕ holds.

Next we prove 2. Assume that **AG AF** ϕ holds and there exists a scc such that when all states satisfying ϕ are removed from scc we still have a cycle consisting of states in scc. In that case, we can find a path $s_0, s_1 \ldots s$ leading to a state s on this cycle, and we can then extend this to an infinite path by repeating the states on the cycle to which s belong. Since no state on the cycle satisfy ϕ, then **AG AF** ϕ cannot hold. Hence, we cannot have such cycles. Assume now that each strongly connected component becomes acyclic when removing states satisfying ϕ. Since all cycles belongs to some strongly connected component, then we cannot have cycles where no states satisfy ϕ. Thus, from any states on an infinite path we must eventually encounter a state satisfying ϕ which means that **AG AF** ϕ holds.

Based on Proposition 2 we can now specify the CHECKPROPERTY procedure which is given in Algorithm 2. The procedure first computes the SCC of the given layer $\mathcal{L}$. Here any algorithm for computing SCC can be used, and we do not specify this further. Based on the SCC and Proposition 2, the procedure then checks whether the property being investigated is violated in which case false is returned and the entire algorithm terminates. At the end of the algorithm (line 18), true is returned in case the property was never violated.

```
1  begin
2  │   SCC ← ComputeSCC(Layer)
3  │   if Φ ≡ AG EF φ then
4  │   │   forall scc ∈ SCC do
5  │   │   │   if isTerminal(scc) ∧ ∀s ∈ scc : ¬φ(s) then
6  │   │   │   │   return false
7  │   │   │   end
8  │   │   end
9  │   end
10 │   if Φ ≡ AG AF φ then
11 │   │   forall scc ∈ SCC do
12 │   │   │   V ← scc \ {s ∈ scc | φ(s)}
13 │   │   │   if hasCycle(V) then
14 │   │   │   │   return false
15 │   │   │   end
16 │   │   end
17 │   end
18 │   return true
19 end
```

Algorithm 2: Checking strongly connected components of current layer

We have not specified the details of the ISTERMINAL and HASCYCLE procedures. The ISTERMINAL procedure can be implemented by checking that all successors of nodes in the scc are contained in the scc. The HASCYCLE procedure can be implemented by, e.g., a depth-first search of the nodes in V.

The completeness of the basic sweep-line algorithm and Proposition 1 ensures that all strongly connected components will eventually have been computed and inspected in Algorithm 2. Furthermore, Algorithm 2 is a direct implementation of the two properties stated in Proposition 2. We therefore have the following theorem concerning the correctness of our algorithm:

Theorem 1. *Let $\mathcal{P} = (O, \sqsubseteq, \psi)$ be a monotonic progress measure, and let $\Phi \equiv$ AG EF ϕ or $\Phi \equiv$ AG AF ϕ. Then Algorithm 1 terminates and Φ is satisfied if and only if the algorithm returns true.*

In Algorithm 2 we have separated the computation of SCC from the checking of the SCC. As an optimisation it is possible to integrate the checking of the properties of a scc into the scc computation algorithm. This could make it possible to check the SCC as they are encountered by the scc-algorithm. As a further optimisation it is also possible to compute the SCC as the layer is being explored and not at the end of exploring a layer. However, for reason of clarity, we have decided to separate the two steps in the formulation of the algorithm.

As the continuation of the work presented in [17], we have implemented Algorithm 1 using the Standard ML language, and integrated it into CPN Tools. This allows us not only to analyse states spaces of models constructed using CPN Tools taking advantage of the sweep-line method, but also to verify the aforementioned behavioural properties. We have also optimised the algorithm, so every time a property is violated or we know that it cannot be further satisfied, the execution stops to save time.

4   The CPN MQTT Model

Our aim is to use the property-specific sweep-line model checking algorithms for CTL from the previous section to verify the key behavioural properties of the CPN model we have developed of the MQTT protocol [19].

MQTT applies topic-based filtering of messages with a topic being part of each published message. An MQTT client can subscribe to a topic to receive messages, publish on a topic, and clients can subscribe to as many topics as they are interested in. As described in [18], an MQTT client can operate as a publisher or as a subscriber, and we use the term client to generally refer to a publisher or a subscriber. The broker [18] is the core of any publish/subscribe protocol and is responsible for keeping track of subscriptions, receiving and filtering messages, deciding to which clients they will be dispatched, and sending them to all subscribed clients. The MQTT protocol delivers application messages according to the three Quality of Service (QoS) levels defined in [2], which are motivated by the typically needs that IoT applications may have in terms of reliable delivery of messages.

4.1   Interaction Overview

MQTT defines five main operations: connect, subscribe, publish, unsubscribe and disconnect. Such operations, except the connect which must be performed a priori by each of the clients who want to participate in the communication, are mutually independent and can be triggered in parallel by the clients and processed by the broker. We have developed the CPN model following modelling patterns that ensure modularity, and thereby encapsulation of both the protocol logic and the behaviour of such operations.

In order to show how the clients and the broker interact, we describe the different actions that clients may carry out by considering an example. Figure 1 shows a sequence diagram for a scenario where two clients connect, perform subscribe, publish and unsubscribe, and finally disconnect from the broker. The protocol interaction is as follows:

1. Client 1 and Client 2 request a connection to the Broker.
2. The Broker sends back a connection acknowledgement (CONNACK) to confirm the establishment of the connection.
3. Client 2 subscribes to topic 1 with a QoS level 1, and the Broker confirms the subscription with a subscribe acknowledgement message.
4. Client 1 publishes on topic 1 with a QoS level 1. The Broker responds with a corresponding publish acknowledgement (PUBACK).
5. The Broker transmits the publish message to Client 2 which is subscribed to the topic.

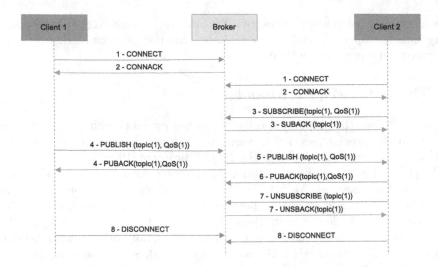

Fig. 1. Message sequence diagram illustrating the MQTT phases.

6. Client 2 gets the published message, and sends a publish acknowledgement back as a confirmation to the Broker that it has received the message.
7. Client 2 unsubscribes to topic 1, and the Broker responds with an unsubscribe acknowledgement.
8. Client 1 and Client 2 disconnect.

4.2 CPN Model Overview

We now briefly show and discuss the model and its main elements that are important for the understanding of the work carried out. We refer the reader to [19] for a detailed description of the MQTT protocol and the MQTT CPN model. The complete CPN model of the MQTT protocol consists of twenty four modules organised into six hierarchical levels.

The model is organised following a modelling pattern that ensures modularity and therefore, encapsulation of the protocol logic and behaviour of such operations. This offers advantages both for readability and understandability of the model and also, for making it easier to detect and fix errors during the incremental verification. For instance, this has allowed us to make a clear separation of the different QoS functional logic without having any negative complexity impact on the model. Note that the verification is incremental in the sense that we start with a core functionality of the protocol, and then we incrementally add more operations until we have the complete functionality included. This implies that we incrementally verify properties associated to each set of the operations.

Figure 2 shows the top-level module of the CPN MQTT model which consists of two *substitution transitions* (drawn as rectangles with double-lined borders) representing the Clients and the Broker roles of MQTT. Substitution transitions constitute the basic syntactical structuring mechanism of CPNs and

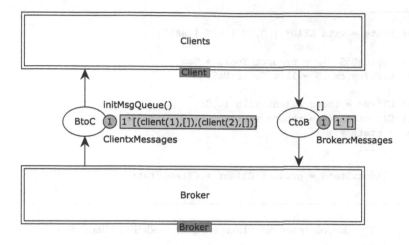

Fig. 2. The top-level module of the MQTT CPN model.

each of the substitution transitions has an associated *module* that models the detailed behaviour of the clients and the broker, respectively. The name of the (sub)module associated with a substitution transition is written in the rectangular tag positioned next to the transition.

The two substitution transitions in Fig. 2 are connected via directed arcs to the two places CtoB and BtoC. The clients and the broker interact by producing and consuming tokens on the places. The places CtoB and BtoC are designed to behave as queues. The queue mechanism offers some advantages that the MQTT specification implicitly indicates. The purpose of this is to ensure the ordered message distribution as assumed from the transport service on top of which MQTT operates.

4.3   Client and Broker State Modelling

The colour sets defined for modelling the client state are shown in Fig. 3. The ClientProcessing submodule in Fig. 4 models all the operations that a client can carry out. Clients can behave as senders and receivers, and the five substitution transitions CONNECT, PUBLISH, SUBSCRIBE, UNSUBSCRIBE and DISCONNECT have been constructed to capture both behaviours.

The place Clients (top-left place in Fig. 4) uses a token for each client to store its respective state during the communication. The State colour set is an enumeration type containing the values READY (for the initial state), WAIT (when the client is waiting to be connected), CON (when the client is connected), and DISC (for when the client has disconnected). The states of the clients are represented by the ClientxState colour set which is a product of Client and ClientState. The colour set ClientState is used to represent the state of a client and consists of a list of TopicxQoS, a State, and a PID. Using this, a client stores the topics it is subscribed to, and the quality of service level of

```
colset State = with READY | DISC | CON | WAIT;

colset TopicxQoS    = product Topic * QoS;
colset ListTopicxQoS = list TopicxQoS;

colset Client = index client with 1..C;
colset ClientState  = record topics : ListTopicxQoS *
state   : State *
pid     : PID;

colset ClientxState = product Client * ClientState;
```

Fig. 3. Colour set definitions used for modelling client state.

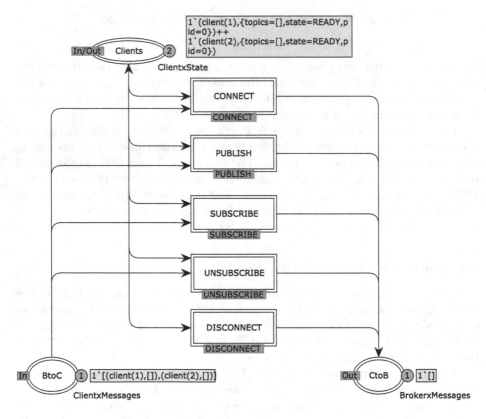

Fig. 4. ClientProcessing submodule.

each subscription. The colour set PID is used for modelling the packet identifiers which play a central role in the MQTT protocol logic.

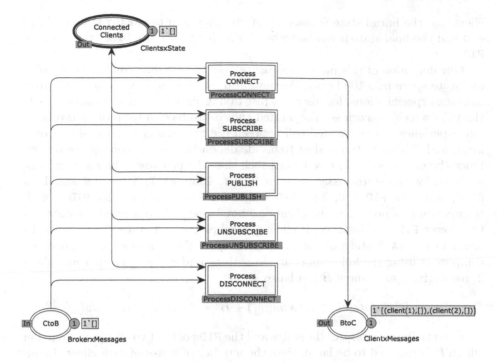

Fig. 5. The BrokerProcessing module.

We have structured the broker similarly as we have done for clients. This can be seen from Fig. 5 which shows the BrokerProcessing submodule. The ConnectedClients place keeps the information of all clients as perceived by the broker. This place is designed as a central storage, and it is used by the broker to distribute the messages over the network. The broker behaviour is different from that of the clients, since it will have to manage all the requests and generate responses for several clients at the same time.

5 Model Checking and Experimental Results

In this section we show how we have performed sweep-line based model checking of the CPN MQTT model and present the results from the experiments.

5.1 Progress Measure

The first aspect to consider is how to define the progress measure of the model. Since the model runs in an acyclic configuration there is a final state where all the clients are disconnected and we take advantage of the PID as a way to keep track of the evolution of the message interchange. We have therefore defined the progress measure as a combination of the different states the clients can go through in conjunction with the PIDs. In the experiments, we consider two clients, so the initial state is made up of two clients in the READY state and PID = 0 and the final state is reached when both clients are in a DISC state and the PID = 3.

Our definition of this progress measure over the possible combinations splits our state space into 100 layers. We have also experimented with other progress measures specifications, for instance, just taking into account the states or only the PIDs which for each such separated choice produces a total of 16 layers. In our experience, there is a trade-off between the granularity and the size of each layer, and it is up to the analyst to decide depending on the concrete resources. Since the progress measure is defined such that the progress values are integers, we have for the states assigned 1 for READY, 2 for WAIT, 3 for CON and 4 for DISC, and 1 for PID = 0, 2 for PID = 1, 3 for PID = 2 and 4 for PID = 4. It is important to note that the clients cannot backtrack to a previous state nor to a lower PID. For instance, if client 1 reaches the CON state, it can never be again in the WAIT state. As we need to keep a global notion of progress, we compute it using the following equation with c_1 and c_2 being client 1 and client 2, respectively and where B is a base:

$$\psi_c = B^3 * state(c_1) + B^2 * pid(c_1) + B^1 * state(c_2) + B^0 * pid(c_2)$$

Essentially, we interpret the states and the PIDs of the two clients as a number where B is required to be larger than the number of states of each client. In our experiments, we have used $B = 10$, i.e., the decimal numbering system. With this, we can obtain a progress value for each possibility (between 1111 and 4444) and respecting the monotonic ordering of non-regress.

As we have implemented the model in a modular and parameterized fashion, we are able to control several elements, for instance, the number of clients, the operations those clients can perform (e.g., connect and subscribe), and the size of the queues for handling messages. Note that, in order to obtain a finite state space, we have to limit the number of clients and topics, and also bound the packet identifiers. The packet identifiers are incremented throughout the execution of the different phases of the protocol, i.e., the connect, subscribe, data exchange, unsubscribe, and disconnect phases. This means that we cannot use a single global bound on the packet identifiers as a client could reach this bound, e.g., already during the publish phase and hence the global bound would prevent (block) a subsequent unsubscribe to take place. We therefore introduce a local upper bound on packet identifiers for each phase. This local bound expresses that the given phase may use packet identifiers up to this local bound. In the next subsection, we present the results of, first, running the state space using the sweep-line algorithm, and second, verifying certain behavioural properties.

5.2 Incremental Verification and Properties

We have designed a system to run six incremental executions which gives us more control to detect errors during the validation of the model and the verification of the properties. The six different scenarios are wrapped within three different steps. In the first step we include only the parts related to clients connecting and disconnecting. In the second step we add subscribe and unsubscribe, and finally in the third step we add data exchange considering the three quality of service levels in turn. At each step, we include verification of additional properties. Below we briefly discuss the three steps and the properties verified at each step. Note that properties that reason about clients are verified for each individual client. In other words, the properties make sure that every client involved satisfies the property being verified.

Step 1. Connect and Disconnect. In this first step we consider only the part of the model related to clients connecting and disconnecting to the broker.

S1-P1-ConsistentConnect. The clients and the broker have a consistent view of the connection state.

S1-P2-ClientsCanConnect. There exists a reachable state in which each client is connected to the broker.

S1-P3-ConsistentTermination. Each terminal state (dead marking) has a consistent and desired behaviour.

S1-P4-PossibleTermination. The protocol can always be terminated, i.e., a terminal state (dead marking) can always be reached.

Step 2. Subscribe and Unsubscribe. In this step, we add the ability for the clients to subscribe and unsubscribe (in addition to connect/disconnect from step 1).

S2-P1-CanSubscribe. There exists states in which both the clients and the broker sides consider each client to be subscribed.

S2-P2-ConsistentSubscription. In every state there is a consistent subscription in both clients and broker sides.

S2-P3-PossiblySubscribed. If the client sends a subscribe message, then eventually both the clients and the broker sides will consider the client to be subscribed.

S2-P4-CanUnsubscribe. For each client there exists executions in which the client sends an unsubscribe message.

S2-P5-EventuallyUnsubscribed. If the client sends an unsubscribe message, then eventually that both the clients and the broker sides consider the client to be unsubscribed.

Step 3. Publish and QoS levels. We add the ability for the clients to publish and receive messages in addition to the rest of the properties of Steps 1 and 2.

S3-P1-PublishConnect. Each client can publish if it is in a connected state.

S3-P2-CanPublish. There exists an execution in which each client publishes a message.

S3-P3-CanReceive. For each client there exists an execution in which each client receives a message.

S3-P4-ReceiveSubscribed. A client only receives data if it is subscribed to the topic, i.e., the client side considers the client to be subscribed.

Table 1 shows the representation of the properties in CTL. Note that the verified properties have the forms described in Sect. 3. We have marked in Table 1 some properties with "*". The property S2-P3 has been computed as if it were an *EF* property (the same applies to S2-P5). However, this does not completely verify the property since it only checks that it is possible to find a state where the client is subscribed. What we really want to check is that we can reach a state where the client sends a subscribe message, and eventually after that the client is subscribed in the broker side. The implementation of such properties of the form $AG(\Phi \Rightarrow AF(\Psi))$ is part of our future work.

5.3   Experimental Results

Table 2 summarises the statistics as a result of running the six scenarios, using both approaches, the traditional CPN state space exploration and the sweep-line method approach, and verifying the properties aforementioned. The States and Arcs columns give the number of states and edges, respectively, in the state space. The Peak column lists the peak number of states stored in memory (i.e., the number of states in the largest layer). The Rel. Mem. Reduction column indicates the reduction of memory as the result of using the sweep-line method, compared to the total number of states (stored in memory by the tradition approach). For instance, in row number 5 in Table 2, we have a reduction in memory consumed of 84.17%, which means that the number of states we have in memory corresponds to the 15.83% of the total amount of states we would store using the traditional approach. The TV-Time column amounts the time that took for the traditional procedure to verify the properties. The SLV-Time column details the time needed to verify the properties using the sweep-line approach. Finally, the column Rel. Time Increment gives the relative additional

Table 1. CTL properties verified.

Property	CTL formula	Description
S1-P1	AGΦ	Φ: Consistent connection
S1-P2	EFΦ	Φ: Each client is connected to the broker
S1-P3	AG(¬ DM ∨ Φ)	DM: Dead marking | Φ: desired dead marking
S1-P4	AGEF DM	DM: Dead marking (checked in S1-P3 that it is desired)
S2-P1	EFΦ	Φ: Each client can subscribe
S2-P2	AGΦ	Φ: Each client is consistently subscribed
S2-P3*	EFΦ	Explanation above
S2-P4	EFΦ	Φ: Each client can unsubscribe
S2-P5*	EFΦ	Explanation above
S3-P1	AG (Φ ⇒ Ψ)	Φ: Client connected | Ψ: Client can publish
S3-P2	EFΦ	Φ: Each client can send a publish
S3-P3	EFΦ	Φ: Each client can receive a publish
S3-P4	AG (Φ ⇒ Ψ)	Φ: Client receives a publish | Ψ: Client is subscribed

Table 2. Results on the six incremental executions using both approaches.

Configuration	States	Arcs	Peak	Rel. Mem. Reduction	TV-Time	SLV-Time	Rel. Time Increment
1. Conn-Disconn	35	48	9	74.29%	0.00 s	0.00 s	0%
2. 1 + Subscribe	507	1,054	180	64.50%	0.156 s	0.219 s	79%
3. 2 + Unsubscribe	1,849	4,120	300	83.78%	1.328 s	2.171 s	63.48%
4. 3 + Pub QoS 0	4,282	8,840	711	83.4%	4.453 s	4.983 s	11.9%
5. 3 + Pub QoS 1	11,462	23,934	1,815	84.17%	20.172 s	28.531 s	41.44%
6. 3 + Pub QoS 2	43,791	85,682	7,037	83.93%	168.113 s	250.708 s	49.13%

time that was necessary for the sweep-line method to proceed, compared to the traditional approach.

The two approaches provided the same results during the evaluation of the properties, keeping the consistency of the verification process. Even though the sweep-line is more time consuming, the memory usage was successfully reduced even in the worst case scenario. The highest relative time consumption is located in the third row with an increase of 63.48%. However, this should not be taken completely as reference since the calculation with such a low number of states and arcs is very sensitive to also the time that takes to compute the state space and the SCC.

6   Conclusions and Future Work

We have presented the application of the sweep-line method for verifying an elaborate set of behavioral properties of the MQTT protocol. The application of the sweep-line method relied on a set of on-the-fly algorithms for model checking selected CTL behavioral properties. We have compared the application of the sweep-line method with the application of standard CTL model checking in CPN Tools demonstrating a substantial reduction in memory usage at the expense of a modest increase in execution time. The consistency between the results obtained using conventional CTL model checking and the results obtained with the implementation of our property-specific CTL model checking algorithms for the sweep-line method serves as a validation of our new approach.

We see several possible directions for future work based on the results and experiments presented in this paper. We plan to investigate a more complete set of scenarios where different configurations are considered. This includes the number of clients, different progress measures, distinct queue sizes, and the possibility of retransmitting packets. This is going to be relevant to make other analysis and study, first, how the number and size of the strongly connected components affects the sweep-line method and second, how the reduction factor grows with the value of the parameter. Related to this, there are also several possibilities for improving the implementation of the property-specific CTL model checking algorithms that we employ.

CTL model checking with the sweep-line method has until now been an open research problem, and the algorithms presented represents a first step towards addressing this. The extension of our approach to cover a larger subset of CTL properties is an important direction of future work. An example is the *S2-P3-EventualSubscribed* property discussed in Sect. 5. Properties on this form can be explored in a two-steps fashion way, where first the property in the left-hand side of the implication is accomplished, and then a second instance of the state space is explored, checking whether the property in the right-hand side is satisfied or not. The work presented in [16] on using tailored model checking algorithms for different CTL properties could serve as a starting point. A key challenge is to identity a subset of CTL compatible with the least-progress-first exploration order of the sweep-line method. In the context of symbolic model checking using binary-decision diagrams (BDDs), forward CTL model checking algorithms have been developed [11]. However, the sweep-line method is not compatible with the use of BDDs. The reason is that deleting states from a BDD (as required by the sweep-line method) may cause the memory usage for storing the BDD to increase. This counteracts the idea of how the sweep-line method alleviates the state explosion problem.

A more open direction of future work is to develop CTL model checking techniques that can be used for non-monotonic progress measures - and not only monotonic progress measures as presented in this paper. We see potential improvements in being capable of including non-monotonic progress measures. It would significantly expand the class of models that can be analysed, for instance, we could also run the algorithm in the cyclic version of the CPN MQTT model.

References

1. Baier, C., Katoen, J.-P.: Principles of Model Checking. MIT Press, Cambridge (2008)
2. Banks, A., Gupta, R.: MQTT Version 3.1.1. OASIS Stand. **29**, 89 (2014). http://docs.oasis-open.org/mqtt/mqtt/v3.1.1/mqtt-v3.1.1.html
3. Cheng, A., Christensen, S., Mortensen, K.H.: Model checking coloured petri nets - exploiting strongly connected components. DAIMI Rep. Ser. **26**, 519 (1997)
4. Christensen, S., Kristensen, L.M., Mailund, T.: A sweep-line method for state space exploration. In: Margaria, T., Yi, W. (eds.) TACAS 2001. LNCS, vol. 2031, pp. 450–464. Springer, Heidelberg (2001). https://doi.org/10.1007/3-540-45319-9_31
5. Clarke, E.M., Emerson, E.A.: Design and synthesis of synchronization skeletons using branching time temporal logic. In: Kozen, D. (ed.) Logic of Programs 1981. LNCS, vol. 131, pp. 52–71. Springer, Heidelberg (1982). https://doi.org/10.1007/BFb0025774
6. Clarke, E.M., Emerson, E.A., Sistla, A.P.: Automatic verification of finite-state concurrent systems using temporal logic specifications. ACM Trans. Program. Lang. Syst. (TOPLAS) **8**(2), 244–263 (1986)
7. Clarke, E.M., Grumberg, O., Minea, M., Peled, D.: State space reduction using partial order techniques. Int. J. Softw. Tools Technol. Transf. **2**(3), 279–287 (1999)

8. Clarke, E.M., Klieber, W., Nováček, M., Zuliani, P.: Model checking and the state explosion problem. In: Meyer, B., Nordio, M. (eds.) LASER 2011. LNCS, vol. 7682, pp. 1–30. Springer, Heidelberg (2012). https://doi.org/10.1007/978-3-642-35746-6_1

9. CPN tools. http://cpntools.org/

10. Evangelista, S., Kristensen, L.M.: Hybrid on-the-fly LTL model checking with the sweep-line method. In: Haddad, S., Pomello, L. (eds.) PETRI NETS 2012. LNCS, vol. 7347, pp. 248–267. Springer, Heidelberg (2012). https://doi.org/10.1007/978-3-642-31131-4_14

11. Iwashita, H., Nakata, T., Hirose, F.: CTL model checking based on forward state traversal. In: Proceedings of International Conference on Computer Aided Design, pp. 82–87. IEEE Computer Society (1996)

12. Jensen, K., Kristensen, L., Mailund, T.: The sweep-line state space exploration method. Theor. Comput. Sci. **429**, 169–179 (2012)

13. Jensen, K., Kristensen, L.M., Wells, L.: Coloured petri nets and CPN tools for modelling and validation of concurrent systems. Int. J. Softw. Tools Technol. Transf. **9**(3), 213–254 (2007)

14. Kristensen, L.M., Mailund, T.: A generalised sweep-line method for safety properties. In: Eriksson, L.-H., Lindsay, P.A. (eds.) FME 2002. LNCS, vol. 2391, pp. 549–567. Springer, Heidelberg (2002). https://doi.org/10.1007/3-540-45614-7_31

15. Kristensen, L.M., Christensen, S.: Implementing coloured petri nets using a functional programming language. Higher-order Symbolic Comput. **17**(3), 207–243 (2004)

16. Liebke, T., Wolf, K.: Taking some burden off an explicit CTL model checker. In: Donatelli, S., Haar, S. (eds.) PETRI NETS 2019. LNCS, vol. 11522, pp. 321–341. Springer, Cham (2019). https://doi.org/10.1007/978-3-030-21571-2_18

17. Lilleskare, A., Kristensen, L.M., Høyland, S.-O.: CTL model checking with the sweep-line state space exploration method. In: Proceedings of Norwegian Informatics Conference (NIK) (2017)

18. MQTT essentials part 3: Client, broker and connection establishment. https://www.hivemq.com/blog/mqtt-essentials-part2-publish-subscribe

19. Rodríguez, A., Kristensen, L.M., Rutle, A.: Formal modelling and incremental verification of the MQTT IoT protocol. In: Koutny, M., Pomello, L., Kristensen, L.M. (eds.) Transactions on Petri Nets and Other Models of Concurrency XIV. LNCS, vol. 11790, pp. 126–145. Springer, Heidelberg (2019). https://doi.org/10.1007/978-3-662-60651-3_5

20. Rodriguez, A., Kristensen, L.M., Rutle, A.: On CTL model checking of the MQTT IoT protocol using the sweep-line method. In: Petri Nets and Software Engineering. International Workshop, PNSE 19, Aachen, Germany, June 24, 2019, volume 2424 of CEUR Workshop Proceedings, pp. 57–72 (2019)

21. Stern, U., Dill, D.L.: Improved probabilistic verification by hash compaction. In: Camurati, P.E., Eveking, H. (eds.) CHARME 1995. LNCS, vol. 987, pp. 206–224. Springer, Heidelberg (1995). https://doi.org/10.1007/3-540-60385-9_13

22. Valmari, A.: The state explosion problem. In: Advanced Course on Petri Nets, pp. 429–528. Springer (1996)

23. Van Leeuwen, J., Leeuwen, J.: Handbook of Theoretical Computer Science, vol. 1. Mit Press, Elsevier (1990)

24. Vardi, M.Y.: Branching vs. Linear time: final showdown. In: Margaria, T., Yi, W. (eds.) TACAS 2001. LNCS, vol. 2031, pp. 1–22. Springer, Heidelberg (2001). https://doi.org/10.1007/3-540-45319-9_1

Correction to: Transactions on Petri Nets and Other Models of Concurrency XV

Maciej Koutny, Fabrice Kordon⊙, and Lucia Pomello

Correction to:
M. Koutny et al. (Eds.): *Transactions on Petri Nets and Other Models of Concurrency XV*, LNCS 12530, https://doi.org/10.1007/978-3-662-63079-2

The original version of this publication was revised. The affiliation of Lucia Pomello was corrected to "Università degli Studi di Milano-Bicocca, Milan, Italy".

The updated version of the book can be found at
https://doi.org/10.1007/978-3-662-63079-2

Author Index

Adobbati, Federica   50, 126
Aubel, Adrián Puerto   50

Bernardinello, Luca   126
Berti, Alessandro   1

de Lara, Juan   27
Devillers, Raymond   75

Erofeev, Evgeny   75

Ferigato, Carlo   50

Gandelli, Stefano   50
Gómez-Martínez, Elena   27
Guerra, Esther   27

Hujsa, Thomas   75

Kristensen, Lars Michael   165

Liebke, Torsten   150

Pomello, Lucia   126

Rodríguez, Alejandro   165
Rutle, Adrian   165

Tredup, Ronny   101

van der Aalst, Wil M. P.   1

Wolf, Karsten   150

Printed in the United States
by Baker & Taylor Publisher Services